Introduction to MINES and MINERALS

Introduction to MINES and MINERALS

Alec Rivera

Oxford Book Company

Jaipur ▪ New Delhi

Published By

Oxford Book Company

Regd. Off. "Bony Residency", Gate No. 2
Opp. Tilak Public School, Vishweswariya Nagar,
Gopalpura Road, Jaipur - 302018, India
Ph: 0141 - 2761280, Telefax: 2761381
email - oxfordbookjpr@gmail.com
website - oxfordbookcompany.com

Delhi Office: 102, 1st Floor, Satyam House,
4327/3, Ansari Road, Darya Ganj,
New Delhi - 110002, India
Ph: 011-45652440

Distributed by
KEYA INTERNATIONAL PRESS LLP
E-mail: keyainternationalpressllp@gmail.com

First published in India, 2024

ISBN : 9789355240705

Typeset by Shivangi Computers, Jaipur

Printed and bound in India by Replika Press Pvt. Ltd.

Contents

Preface *vii*

1. **Introduction** **1**
Mineral • Crystal Structure • Chemical Properties of Minerals

2. **Native Element Minerals** **13**
Sulfosalt Minerals • Halide Minerals • Fluorite Structure • Oxide Minerals • Arsenite Minerals • Carbonate Minerals • Borate Minerals • Sulfate Minerals • Phosphate Minerals

3. **List of Minerals** **64**
Abenakiite-(Ce) • Abernathyite • Abhurite • Abswurmbachite • Acanthite • Actinolite • Acuminite • Adamite • Adamsite-(Y) • Adelite • Admontite • Aegirine • Aenigmatite • Aerinite • Aerugite • Aeschynite-(Ce) • Aeschynite-(Nd) • Aeschynite-(Y) • Abelsonite • Afghanite • Afwillite • Agardite • Ajoite • Akaganéite • Agate • Alabaster • Chrysoberyl • Amber • Alum

4. **Important Minerals** **109**
Amazonite • Amethyst • Ammolite • Grunerite • Anyolite • Antozonite • Beryl • Argentite • Asbestos • Contamination of Other Products • Asbestos in Construction • Critics of Safety Regulations • Illite • Aventurine • Aksaite • Alabandite • Alamosite • Alforsite • Aliettite

5. **Crystal and Ores** **157**
Allabogdanite • Allanite • Alloclasite • Almandine • Alstonite • Altaite • Aluminite • Aluminium • Alunite • Alunogen • Amblygonite • Ameghinite • Amphibole • Analcime • Anapaite • Anatase • Andalusite • Andesine • Andradite • Anglesite

• Anhydrite • Ankerite • Annabergite • Anorthite • Anorthoclase • Anthophyllite

6. Serpentine Group **196**
Antimony • Antitaenite • Antlerite • Barium • Beryllium • Bismuth • Boron • Cement • Types of Modern Cement

Bibliography 259

Index 263

Preface

A mineral is a naturally occurring solid with a definite chemical composition and a specific crystalline structure. A rock is an aggregate of one or more minerals. Some rocks are predominantly composed of just one mineral. For example, limestone is a sedimentary rock composed almost entirely of the mineral calcite. Other rocks contain many minerals, and the specific minerals in a rock can vary widely. Some minerals, like quartz, mica or feldspar are common, while others have been found in only four or five locations worldwide. The vast majority of the rocks of the Earth's crust consist of quartz, feldspar, mica, chlorite, kaolin, calcite, epidote, olivine, augite, hornblende, magnetite, hematite, limonite and a few other minerals. Over half of the mineral species known are so rare that they have only been found in a handful of samples, and many are known from only one or two small grains. Commercially valuable minerals and rocks are referred to as industrial minerals. Rocks from which minerals are mined for economic purposes are referred to as ores.

A main determining factor in the formation of minerals in a rock mass is the chemical composition of the mass, for a certain mineral can be formed only when the necessary elements are present in the rock. Calcite is most common in limestones, as these consist essentially of calcium carbonate; quartz is common in sandstones and in certain igneous rocks which contain a high percentage of silica. Other factors are of equal importance in determining the natural association or paragenesis of rock-forming minerals, principally the mode of origin of the rock and the stages through which it has passed in attaining its present condition. Two rock masses may have very much the same bulk composition and yet consist of entirely different assemblages of minerals. The tendency is always for those compounds to be formed which are stable under the conditions under which the rock mass originated. A granite arises by the consolidation of a molten magma at high temperatures and great pressures and its component minerals are those stable under such conditions.

Exposed to moisture, carbonic acid and other subaerial agents at the ordinary temperatures of the Earth's surface, some of these original minerals, such as quartz and white mica are relatively stable and remain unaffected; others weather or decay and are replaced by new combinations. The feldspar passes into kaolinite, muscovite and quartz, and any mafic minerals such as pyroxenes, amphiboles or biotite have been present they are often altered to chlorite, epidote, rutile and other substances. These changes are accompanied by disintegration, and the rock falls into a loose, incoherent, earthy mass which may be regarded as a sand or soil. The materials thus formed may be washed away and deposited as sandstone or siltstone. The structure of the original rock is now replaced by a new one; the mineralogical constitution is profoundly altered; but the bulk chemical composition may not be very different.

The book incorporates salient features of the earth—its structure and composition.

— Alec Rivera

Chapter 1

Introduction

Mineral

A mineral is a naturally occurring solid chemical substance formed through biogeochemical processes, having characteristic chemical composition, highly ordered atomic structure, and specific physical properties. By comparison, a rock is an aggregate of minerals and/or mineraloids and does not have a specific chemical composition. Minerals range in composition from pure elements and simple salts to very complex silicates with thousands of known forms. The study of minerals is called mineralogy.

Mineral Definition and Classification

To be classified as a true mineral, a substance must be a solid and have a crystalline structure. It must also be a naturally occurring, homogeneous substance with a defined chemical composition.

The International Mineralogical Association approved the following definition in 1995:

> *A mineral is an element or chemical compound that is normally crystalline and that has been formed as a result of geological processes.f*

According to this definition and classification scheme, biogenic materials were excluded from the mineral kingdom:

> *Biogenic substances are chemical compounds produced entirely by biological processes without a geological component (e.g., urinary calculi, oxalate crystals in plant tissues, shells of marine molluscs, etc.) and are not regarded as minerals. However, if geological processes were involved in the genesis of the compound, then the product can be accepted as a mineral.f*

However, other researchers do not adhere to this exclusion rule. Lowenstam (1981), for example, states the following:

> *Organisms are capable of forming a diverse array of minerals, some of which cannot be formed inorganically in the biosphere.f*

The distinction is a matter of classification and less to do with the constituents of the minerals themselves. Skinner (2005) views all solids as potential minerals and includes biominerals in the mineral kingdom, which are those that are created by the metabolic activities of organisms. Inclusion of these biogenic minerals requires a expanded definition of a mineral as:

> *An element or compound, amorphous or crystalline, formed through* biogeochemical *processes.f*

Mineral classification schemes and their definitions are evolving to match recent advances in mineral science. More recent classifications, for example, include an *organic class* – in both the new Dana and the Strunz classification schemes. The organic class includes a very rare group of minerals with hydrocarbons.

The IMA Commission on New Minerals and Mineral Names recently adopted (in 2009) a hierarchical scheme for the naming and classification of mineral groups and group names and established seven commissions and four working groups to review and classify minerals into an official listing of their published names. According to these new rules, "mineral species can be grouped in a number of different ways, on the basis of chemistry, crystal structure, occurrence, association, genetic history, or resource, for example, depending on the purpose to be served by the classification."

Recent advances in high-resolution genetic and x-ray absorption spectroscopy is opening new revelations on the biogeochemical relations between microrganisms and minerals that may make Nickel's (1995) biogenic mineral exclusion obsolete and Skinner's (2005) biogenic mineral inclusion a necessity. For example, the IMA commissioned 'Environmental Mineralogy and Geochemistry Working Group' deals with minerals in the hydrosphere, atmosphere, and biosphere. Mineral forming microorganisms inhabit the areas that this working group deals with. These organisms exist on nearly every rock, soil, and particle surface spanning the globe reaching depths at 1600 meters below the sea floor (possibly further) and 70 kilometres into the stratosphere (possibly entering the mesosphere). Biologists and geologists have recently started to research and appreciate the magnitude of mineral geoengineering that these creatures are capable of. Bacteria have contributed to the formation of minerals for billions of years and critically define the biogeochemical cycles on this planet. Microorganisms can precipitate metals from solution contributing to the formation of ore deposits in addition to their ability to catalyse mineral dissolution, to respire, precipitate, and form minerals.

Prior to the International Mineralogical Association's listing, over 60 biominerals had been discovered, named, and published. These minerals (a sub-set tabulated in Lowenstam (1981)) are considered minerals proper according to the Skinner (2005) definition. These biominerals are not listed in the International Mineral Association official list of mineral names, however, many of these biomineral representatives are distributed amongst the 78 mineral classes listed in the 'Dana' classification scheme. Another rare class of minerals (primarily biological in origin) include the mineral liquid crystals that are crystalline and liquid at the same time. To date over 80,000 liquid crystaline compounds have been identified.

Concerning the use of the term "mineral" to name this family of liquid crystals, one can argue that the term inorganic would be more appropriate. However, inorganic liquid crystals have long been used for organometallic liquid crystals. Therefore in order to avoid any confusion between these fairly chemically different families, and taking into account that a large number of these liquid crystals occur naturally in nature, we think that the use of the old fashioned but adequate "mineral" adjective taken sensus largo is more specific that an alternative such as "purely inorganic", to name this subclass of the inorganic liquid crystals family.

The Skinner (2005) definition of a mineral takes this matter into account by stating that a mineral can be crystalline *or* amorphous. Liquid mineral crystals are amorphous. Biominerals and liquid mineral crystals, however, are not the primary form of minerals, most are geological in origin, but these groups do help to identify at the margins of what constitutes a mineral proper.

Crystal Structure

A *crystal structure* is the orderly geometric spatial arrangement of atoms in the internal structure of a mineral. There are 14 basic crystal lattice arrangements of atoms in three dimensions, and these are referred to as the 14 "Bravais lattices". Each of these lattices can be classified into one of the seven crystal systems, and all crystal structures currently recognized fit in one Bravais lattice and one crystal system.

This crystal structure is based on regular internal atomic or ionic arrangement that is often expressed in the geometric form that the crystal takes. Even when the mineral grains are too small to see or are irregularly shaped, the underlying crystal structure is always periodic and can be determined by X-ray diffraction.

Chemistry and crystal structure together define a mineral. In fact, two or more minerals may have the same chemical composition, but differ in crystal structure (these are known as *polymorphs*).

For example, pyrite and marcasite are both iron sulfide, but their arrangement of atoms differs. Similarly, some minerals have different chemical compositions, but the same crystal structure: for example, halite (made from sodium and chlorine), galena (made from lead and sulfur) and periclase (made from magnesium and oxygen) all share the same cubic crystal structure.

Crystal structure greatly influences a mineral's physical properties. For example, though diamond and graphite have the same composition (both are pure carbon), graphite is very soft, while diamond is the hardest of all known minerals.

This happens because the carbon atoms in graphite are arranged into sheets which can slide easily past each other, while the carbon atoms in diamond form a strong, interlocking three-dimensional network.

There are currently more than 4,000 known minerals, according to the International Mineralogical Association (IMA), which is responsible for the approval of and naming of new mineral species found in nature. Of these, perhaps 100 can be called "common", 50 are "occasional", and the rest are "rare" to "extremely rare".

Mineral Groups and Solid Solution

The chemical composition may vary between end members of a mineral system.

For example the plagioclase feldspars comprise a continuous series from sodium and silicon-rich albite ($NaAlSi_3O_8$) to calcium and aluminium-rich anorthite ($CaAl_2Si_2O_8$) with four recognized intermediate compositions between. Mineral-like substances that don't strictly meet the definition are sometimes classified as mineraloids.

Minerals with the same structure and forming solid solutions are named isomorphs, and form series; for example: forsterite and fayalite of the olivine series, ferberite and hubnerite of the wolframite series. Minerals with the same structure and not forming solid solutions are named isotypes, and form groups [classification of minerals (non silicates)]. Minerals with a similar structure are grouped in homeotype families: amphibole and pyroxene families [classification of minerals (silicates)].

Some ion groups with a similar radius can occupy the same structural site in the crystal cell:

- O^{2-} and OH^- with 1.32 and 1.33 Å respectively.
- Si^{4+} and Al^{3+} with 0.42 and 0.51 Å respectively, the charge is neutralized through an exchange of the other cations:
    - o Si^{4+} – (Al^{3+} and Na^+) or (Si^{4+} and Na^+) – (Al^{3+} and Ca^{2+}).
- Larger molecules may have an unoccupied structural site by occupying another unoccupied structural site or by using a divalent cation instead of two monovalent cations, for instance (amphibole family).
- By the end of the 18th century, the minerals were getting chemical formulas. There were some difficulties, as elements were being discovered and isolated for the first time. The minerals: gadolinite-(Y), aeschynite-(Ce) (first publication: 1830, 04.DF.05), vanadinite, aenigmatite, labyrinthite (IMA 2002-065, 09.CO.10), illustrate these difficulties.

More recent definitions:

- "A mineral group consists of two or more minerals with the same (isotypic) or essentially the same (homeotypic) structure, and composed of chemically similar elements" (IMA-CNMNC).
- "two structures are considered homeotypic if all essential features of topology are preserved between them" (IUCr).

Differences between Minerals and Rocks

A mineral is a naturally occurring solid with a definite chemical composition and a specific crystalline structure. A rock is an aggregate of one or more minerals. (A rock may also include organic remains and mineraloids.) Some rocks are predominantly composed of just one mineral. For example, limestone is a sedimentary rock composed almost entirely of the mineral calcite. Other rocks contain many minerals, and the specific minerals in a rock can vary widely. Some minerals, like quartz, mica or feldspar are common, while others have been found in only four or five locations worldwide.

The vast majority of the rocks of the Earth's crust consist of quartz, feldspar, mica, chlorite, kaolin, calcite, epidote, olivine, augite, hornblende, magnetite, hematite, limonite and a few other minerals. Over half of the mineral species known are so rare that they have only been found in a handful of samples, and many are known from only one or two small grains.

Commercially valuable minerals and rocks are referred to as industrial minerals. Rocks from which minerals are mined for economic purposes are referred to as ores (the rocks and minerals that remain, after the desired mineral has been separated from the ore, are referred to as tailings).

Mineral Composition of Rocks

A main determining factor in the formation of minerals in a rock mass is the chemical composition of the mass, for a certain mineral can be formed only when the necessary elements are present in the rock. Calcite is most common in limestones, as these consist essentially of calcium carbonate; quartz is common in sandstones and in certain igneous rocks which contain a high percentage of silica.

Other factors are of equal importance in determining the natural association or paragenesis of rock-forming minerals, principally the mode of origin of the rock and the stages through which it has passed in attaining its present condition. Two rock masses may have very much the same bulk composition and yet consist of entirely different assemblages of minerals.

The tendency is always for those compounds to be formed which are stable under the conditions under which the rock mass originated. A granite arises by the consolidation of a molten magma at high temperatures and great pressures and its component minerals are those stable under such conditions.

Exposed to moisture, carbonic acid and other subaerial agents at the ordinary temperatures of the Earth's surface, some of these original minerals, such as quartz and white mica are relatively stable and remain unaffected; others weather or decay and are replaced by new combinations.

The feldspar passes into kaolinite, muscovite and quartz, and any mafic minerals such as pyroxenes, amphiboles or biotite have been present they are often altered to chlorite, epidote, rutile and other substances. These changes are accompanied by disintegration, and the rock falls into a loose, incoherent, earthy mass which may be regarded as a sand or soil. The materials thus formed may be washed away and deposited as sandstone or siltstone. The structure of the original rock is now replaced by a new one; the mineralogical constitution is profoundly altered; but the bulk chemical composition may not be very different.

The sedimentary rock may again undergo metamorphism. If penetrated by igneous rocks it may be recrystallized or, if subjected to enormous pressures with heat and movement during mountain building, it may be converted into a gneiss not very different in mineralogical composition though radically different in structure to the granite which was its original state.

Physical Properties of Minerals

Classifying minerals can range from simple to very difficult. A mineral can be identified by several physical properties, some of them being sufficient for full identification without equivocation. In other cases, minerals can only be classified by more complex optical, chemical or X-ray diffraction analysis; these methods, however, can be costly and time-consuming.

Physical properties commonly used are:

- *Crystal structure and habit*: A mineral may show good crystal habit or form, or it may be massive, granular or compact with only microscopically visible crystals.
- *Hardness*: the physical hardness of a mineral is usually measured according to the Mohs scale. This scale is relative and goes from 1 to 10. Minerals with a given Mohs hardness can scratch the surface of any mineral that has a lower hardness than itself.
    - o Mohs hardness scale:
        1. Talc $Mg_3Si_4O_{10}(OH)_2$
        2. Gypsum $CaSO_4{\cdot}2H_2O$
        3. Calcite $CaCO_3$
        4. Fluorite CaF_2
        5. Apatite $Ca_5(PO_4)_3(OH,Cl,F)$
        6. Orthoclase $KAlSi_3O_8$
        7. Quartz SiO_2

8. Topaz $Al_2SiO_4(OH,F)_2$
9. Corundum Al_2O_3
10. Diamond C (pure carbon)

- *Luster* indicates the way a mineral's surface interacts with light and can range from dull to glassy (vitreous).
  - o Metallic – high reflectivity like metal: galena and pyrite
  - o Sub-metallic – slightly less than metallic reflectivity: magnetite
  - o Non-metallic lusters:
    - * *Adamantine* – brilliant, the luster of diamond also cerussite and anglesite
    - * *Vitreous* – the luster of a broken glass: quartz
    - * *Pearly* – iridescent and pearl-like: talc and apophyllite
    - * *Resinous* – the luster of resin: sphalerite and sulfur
    - * *Silky* – a soft light shown by fibrous materials: gypsum and chrysotile
    - * Dull/earthy – shown by finely crystallized minerals: the kidney ore variety of hematite
- *Diaphaneity* describes how well light passes through a mineral; there are three basic degrees of transparency:
  - o *Transparent* objects can be seen through a transparent mineral, such as a clear quartz crystal
  - o *Translucent* light passes through the mineral but no objects can be seen
  - o *Opaque* no light passes through the mineral.

Many minerals range from transparent to translucent or translucent to opaque. Calcite, for instance, can be translucent or opaque. Some minerals that are naturally translucent become opaque with weathering.

- *Color* indicates the appearance of the mineral in reflected light or transmitted light for translucent minerals (i.e. what it looks like to the naked eye).
  - o *Iridescence* – the *play of colours* due to surface or internal interference. Labradorite exhibits internal iridescence whereas hematite and sphalerite often show the surface effect.

- *Streak* refers to the colour of the powder a mineral leaves after rubbing it on an unleased porcelain *streak plate*. Note that this is not always the same colour as the original mineral.
- *Cleavage* describes the way a mineral may split apart along various planes. In thin sections, cleavage is visible as thin parallel lines across a mineral.
- *Fracture* describes how a mineral breaks when broken contrary to its natural cleavage planes.
    - o *Chonchoidal fracture* is a smooth curved fracture with concentric ridges of the type shown by glass.
    - o *Hackley* is jagged fracture with sharp edges.
    - o *Fibrous*
    - o *Irregular*
- *Specific gravity* relates the mineral mass to the mass of an equal volume of water, namely the density of the material. While most minerals, including all the common rock-forming minerals, have a specific gravity of 2.5–3.5, a few are noticeably more or less dense, e.g. several sulfide minerals have high specific gravity compared to the common rock-forming minerals.
- Other properties: fluorescence (response to ultraviolet light), magnetism, radioactivity, tenacity (response to mechanical induced changes of shape or form), piezoelectricity and reactivity to dilute acids.

Chemical Properties of Minerals

Minerals may be classified according to chemical composition. They are here categorized by anion group. The list below is in approximate order of their abundance in the Earth's crust. The list follows the Dana classification system which closely parallels the Strunz classification.

Silicate Class

The largest group of minerals by far are the silicates (most rocks are $\geq$ 95% silicates), which are composed largely of silicon and oxygen, with the addition of ions such as aluminium, magnesium, iron, and calcium. Some important rock-forming silicates include the feldspars, quartz, olivines, pyroxenes, amphiboles, garnets, and micas.

Carbonate Class

The carbonate minerals consist of those minerals containing the anion $(CO_3)^{2-}$ and include calcite and aragonite (both calcium carbonate), dolomite (magnesium/calcium carbonate) and siderite (iron carbonate). Carbonates are commonly deposited in marine settings when the shells of dead planktonic life settle and accumulate on the sea floor. Carbonates are also found in evaporitic settings (e.g. the Great Salt Lake, Utah) and also in karst regions, where the dissolution and reprecipitation of carbonates leads to the formation of caves, stalactites and stalagmites. The carbonate class also includes the nitrate and borate minerals.

Sulfate Class

Sulfate minerals all contain the sulfate anion, SO_4^{2-}. Sulfates commonly form in evaporitic settings where highly saline waters slowly evaporate, allowing the formation of both sulfates and halides at the water-sediment interface.

Sulfates also occur in hydrothermal vein systems as gangue minerals along with sulfide ore minerals. Another occurrence is as secondary oxidation products of original sulfide minerals. Common sulfates include anhydrite (calcium sulfate), celestine (strontium sulfate), barite (barium sulfate), and gypsum (hydrated calcium sulfate). The sulfate class also includes the chromate, molybdate, selenate, sulfite, tellurate, and tungstate minerals.

Halide Class

Figure 1 : Halite

The halide minerals are the group of minerals forming the natural salts and include fluorite (calcium fluoride), halite (sodium chloride), sylvite (potassium chloride), and sal ammoniac (ammonium chloride). Halides, like sulfates, are commonly found in evaporite settings such as salt lakes and landlocked seas such as the Dead Sea and Great Salt Lake. The halide class includes the fluoride, chloride, bromide and iodide minerals.

Oxide Class

Oxide minerals are extremely important in mining as they form many of the ores from which valuable metals can be extracted. They also carry the best record of changes in the Earth's magnetic field. They commonly occur as precipitates close to the Earth's surface, oxidation products of other minerals in the near surface weathering zone, and as accessory minerals in igneous rocks of the crust and mantle.

Common oxides include hematite (iron oxide), magnetite (iron oxide), chromite (iron chromium oxide), spinel (magnesium aluminium oxide – a common component of the mantle), ilmenite (iron titanium oxide), rutile (titanium dioxide), and ice (hydrogen oxide). The oxide class includes the oxide and the hydroxide minerals.

Sulfide Class

Many sulfide minerals are economically important as metal ores. Common sulfides include pyrite (iron sulfide – commonly known as *fools gold*), chalcopyrite (copper iron sulfide), pentlandite (nickel iron sulfide), and galena (lead sulfide). The sulfide class also includes the selenides, the tellurides, the arsenides, the antimonides, the bismuthinides, and the sulfosalts (sulfur and a second anion such as arsenic).

Phosphate Class

The phosphate mineral group actually includes any mineral with a tetrahedral unit AO_4 where A can be phosphorus, antimony, arsenic or vanadium. By far the most common phosphate is apatite which is an important biological mineral found in teeth and bones of many animals. The phosphate class includes the phosphate, arsenate, vanadate, and antimonate minerals.

Element Class

The elemental group includes native metals and intermetallic elements (gold, silver, copper), semi-metals and non-metals (antimony, bismuth, graphite, sulfur). This group also includes natural alloys, such as electrum (a natural alloy of gold and silver), phosphides, silicides, nitrides and carbides (which are usually only found naturally in a few rare meteorites).

Organic Class

The organic mineral class includes biogenic substances in which geological processes have been a part of the genesis or origin of the existing compound. Minerals of the organic class include various oxalates, mellitates, citrates, cyanates, acetates, formates, hydrocarbons and other miscellaneous species. Examples include whewellite, moolooite, mellite, fichtelite, carpathite, evenkite and abelsonite.

Chapter 2

Native Element Minerals

Native element minerals are those elements that occur in nature in uncombined form with a distinct mineral structure. The elemental class includes metals and intermetallic elements, semi-metals and non-metals. This group also includes natural alloys, phosphides, silicides, nitrides and carbides.

There are 32 elements (18 metals, 5 nonmetals and gases) that can be found in nature in their elemental form. Metals give a metallic luster on their polished sides, some nonmetal and semimetal minerals too.

Nickel˘Strunz Classification -01-Native Elements

IMA-CNMNC proposes a new hierarchical scheme (Mills et al., 2009). This list uses the Classification of Nickel–Strunz (10 ed, pending publication).

- Abbreviations:
  - "*" -discredited (IMA/CNMNC status).
  - "?" -questionable/doubtful (IMA/CNMNC status).
  - "REE" -Rare-earth element (Sc, Y, La, Ce, Pr, Nd, Pm, Sm, Eu, Gd, Tb, Dy, Ho, Er, Tm, Yb, Lu)
  - "PGE" -Platinum-group element (Ru, Rh, Pd, Os, Ir, Pt)
  - 03.C Aluminofluorides, 06 Borates, 08 Vanadates (04.H V Vanadates), 09 Silicates:
    - Neso: insular
    - Soro: grouping
    - Cyclo: ring
    - Ino: chain

    * Phyllo: sheet
    * Tekto: three-dimensional framework
- Nickel–Strunz code scheme: NN.XY.##x
  - o NN: Nickel–Strunz mineral class number
  - o X: Nickel–Strunz mineral division letter
  - o Y: Nickel–Strunz mineral family letter
  - o ##x: Nickel–Strunz mineral/group number, x add-on letter.

Class: native Elements

- 01.A Metals and Intermetallic Alloys
  - o 01.AA Copper-cupalite family: 05 Native copper, 05 Lead, 05 Native gold, 05 Native silver, 05 Nickel, 05 Aluminum; 10a Auricupride, 10b Tetra-auricupride; 15 Novodneprite, 15 Khatyrkite, 15 Anyuiite; 20 Cupalite, 25 Hunchunite
  - o 01.AB Zinc-brass family (Cu-Zn alloys): 05 Cadmium, 05 Zinc, 05 Titanium*, 05 Rhenium*; 10a Brass*, 10a Zhanghengite, 10b Danbaite, 10b Tongxinite*
  - o 01.AC Indium-tin family: 05 Indium, 10 Tin; 15 Yuanjiangite, 15 Sorosite
  - o 01.AD Mercury-amalgam family: 00 Amalgam*, 05 Mercury; 10 Belendorffite, 10 Kolymite; 15a Paraschachnerite, 15a Schachnerite, 15b Luanheite, 15c Eugenite, 15d Moschellandsbergite; 20a Weishanite, 20b Goldamalgam*; 25 Potarite, 30 Leadamalgam
  - o 01.AE Iron-chromium family: 05 Kamacite?, 05 Iron, 05 Chromium; 10 Antitaenite*, 10 Taenite, 10 Tetrataenite; 15 Chromferide, 15 Wairauite, 15 Ferchromide; 20 Awaruite, 25 Jedwabite
  - o 01.AF Platinum-group elements: 05 Osmium, 05 Rutheniridosmine, 05 Ruthenium; 10 Palladium, 10 Iridium, 10 Rhodium, 10 Platinum
  - o 01.AG PGE-metal alloys: 05 IMA2008-055, 05 Hexaferrum; 10 Atokite, 10 Zvyagintsevite, 10 Rustenburgite; 15 Taimyrite, 15 Tatyanaite; 20 Paolovite; 25 Plumbopalladinite, 25 Stannopalladinite; 30 Cabriite; 35 Chengdeite, 35 Isoferroplatinum; 40 Ferronickelplatinum, 40 Tetraferroplatinum, 40 Tulameenite; 45 Hongshiite*, 45 Skaergaardite; 50 Yixunite, 55 Damiaoite, 60 Niggliite, 65 Bortnikovite, 70 Nielsenite

- 01.B Metallic Carbides, Silicides, Nitrides and Phosphides
  - o 01.BA Carbides: 05 Cohenite; 10 Isovite, 10 Haxonite; 15 Tongbaite; 20 Khamrabaevite, 20 Niobocarbide, 20 Tantalcarbide; 25 Qusongite, 30 Yarlongite
  - o 01.BB Silicides: IMA2007-036; 05 Mavlyanovite, 05 Suessite; 10 Perryite, 15 Fersilicite*, 20 Ferdisilicite*, 25 Luobusaite, 30 Gupeiite, 35 Hapkeite, 40 Xifengite
  - o 01.BC Nitrides: 05 Roaldite, 10 Siderazot, 15 Carlsbergite, 15 Osbornite
  - o 01.BD Phosphides: 05 Schreibersite, 05 Nickelphosphide; 10 Barringerite, 10 Monipite; 15 Allabogdanite, 15 Florenskyite, 15 Andreyivanovite; 20 Melliniite
- 01.C Metalloids and Nonmetals
  - o 01.CA Arsenic group elements: 05 Bismuth, 05 Antimony, 05 Arsenic, 05 Stibarsen; 10 Arsenolamprite, 10 Pararsenolamprite; 15 Paradocrasite
  - o 01.CB Carbon-silicon family: 05a Graphite, 05b Chaoite, 05c Fullerite; 10a Diamond, 10b Lonsdaleite, 15 Silicon
  - o 01.CC Sulfur-selenium-iodine: 05 Sulfur, 05 Rosickyite; 10 Tellurium, 10 Selenium
- 01.D Nonmetallic Carbides and Nitrides
  - o 01.DA Nonmetallic carbides: 05 Moissanite
  - o 01.DB Nonmetallic nitrides: 05 Nierite, 10 Sinoite
- 01.X Unclassified Strunz Elements (Metals and intermetallic alloys; metalloids and nonmetals; carbides, silicides, nitrides, phosphides)
  - o 01.XX Unknown: 00 Hexamolybdenum, 00 Tantalum*, 00 Brownleeite.

Sulfide minerals

The sulfide minerals are a class of minerals containing sulfide (S^{2+}) as the major anion. Some sulfide minerals are economically important as metal ores. The sulfide class also includes the selenides, the tellurides, the arsenides, the antimonides, the bismuthinides, the sulfarsenides and the sulfosalts. Sulfide minerals are inorganic compounds.

Common or important examples include:

Sulfides:

- Acanthite Ag_2S, Chalcocite Cu_2S, Bornite Cu_5FeS_4, Galena PbS, Sphalerite ZnS, Chalcopyrite $CuFeS_2$, Pyrrhotite $Fe_{1-x}S$, Millerite NiS, Pentlandite $(Fe,Ni)_9S_8$, Covellite CuS, Cinnabar HgS, Realgar AsS, Orpiment As_2S_3, Stibnite Sb_2S_3, Pyrite FeS_2, Marcasite FeS_2, Molybdenite MoS_2

Arsenides:

- Nickeline NiAs
- Skutterudite $(Co,Ni)As_3$.

Antimonides:

- Breithauptite NiSb.

Tellurides:

- Altaite PbTe
- Calaverite $AuTe_2$
- Sylvanite $(Au,Ag)Te_2$.

Sulfarsenides:

- Cobaltite (Co,Fe)AsS
- Arsenopyrite FeAsS
- Gersdorffite NiAsS.

Sulfosalts:

- Pyrargyrite Ag_3SbS_3, Proustite Ag_3AsS_3, Tetrahedrite $Cu_{12}Sb_4S_{13}$, Tennantite $Cu_{12}As_4S_{13}$, Enargite Cu_3AsS_4, Bournonite $PbCuSbS_3$, Jamesonite $Pb_4FeSb_6S_{14}$, Cylindrite $Pb_3Sn_4FeSb_2S_{14}$.

Nickel˘Strunz Classification -02-Sulfides

IMA-CNMNC proposes a new hierarchical scheme (Mills et al., 2009). This list uses the Classification of Nickel–Strunz (10 ed, pending publication).

- Abbreviations:
    - o "*"-discredited (IMA/CNMNC status).
    - o "?"-questionable/doubtful (IMA/CNMNC status).
    - o "REE"-Rare-earth element (Sc, Y, La, Ce, Pr, Nd, Pm, Sm, Eu, Gd, Tb, Dy, Ho, Er, Tm, Yb, Lu)
    - o "PGE"-Platinum-group element (Ru, Rh, Pd, Os, Ir, Pt)
    - o 03.C Aluminofluorides, 06 Borates, 08 Vanadates (04.H V Vanadates), 09 Silicates:

    * Neso: insular
    * Soro: grouping
    * Cyclo: ring
    * Ino: chain
    * Phyllo: sheet
    * Tekto: three-dimensional framework

- Nickel–Strunz code scheme: NN.XY.##x
  - NN: Nickel–Strunz mineral class number
  - X: Nickel–Strunz mineral division letter
  - Y: Nickel–Strunz mineral family letter
  - ##x: Nickel–Strunz mineral/group number, x add-on letter.

Class: sulfides, Selenides, Tellurides

- 02.A Simple Sulfides, Selenides, etc
  - 02.AA Alloys of metalloids with Cu, Ag, Sn, Au: 10a Algodonite, 10b Domeykite, 10d Koutekite; 15 Novakite, 20 Cuprostibite, 25 Kutinaite, 30 Allargentum, 35 Dyscrasite, 40 Maldonite, 45 Stistaite
  - 02.AB Ni-metalloid alloys: 10 Orcelite, 15 Maucherite, 20 Oregonite
  - 02.AC Alloys of metalloids with PGE: 05a Atheneite, 05a Vincentite; 10a Stillwaterite, 10b Mertieite-II, 10c Arsenopalladinite; 15a Miessiite, 15a Isomertieite, 15b Mertieite-I; 20a Stibiopalladinite, 20b Palarstanide, 20c Menshikovite; 25a Palladoarsenide, 25b Rhodarsenide, 25c Palladodymite, 25d Naldretteite, 25e Majakite, 25f Palladobismutharsenide; 30 Polkanovite; 35a Genkinite, 35b Ungavaite, 40 Polarite; 45a Froodite, 45b Iridarsenite, 45c Borishanskiite
- 02.B Metal Sulfides, M:S > 1:1 (mainly 2:1)
  - 02.BA With Cu, Ag, Au: 05a Chalcocite, 05b Djurleite, 05c Geerite, 05d Roxbyite, 05e Digenite, 05f Anilite; 10 Bornite; 15a Berzelianite, 15b Bellidoite, 15c Umangite, 15d Athabascaite; 20a Rickardite, 20b Weissite; 25a Stromeyerite, 25b Mckinstryite, 25c Selenojalpaite, 25c Jalpaite, 25d Eucairite, 25e Henryite; 30a Acanthite, 30a Argentite*, 30b Aguilarite, 30b Naumannite, 30c Hessite, 30d Cervelleite, 30e Stutzite; 35 Argyrodite, 35 Putzite, 35

Canfieldite; 40a Fischesserite, 40a Petzite, 40b Uytenbogaardtite, 40c Petrovskaite, 40d Penzhinite; 45 Bezsmertnovite, 50 Bogdanovite, 55 Bilibinskite, 60 Chenguodaite

- o 02.BB With Ni, Fe: 05 Heazlewoodite; 10 Arsenohauchecornite, 10 Bismutohauchecornite, 10 Hauchecornite, 10 Tellurohauchecornite, 10 Tucekite; 15a Argentopentlandite, 15a Cobaltpentlandite, 15a Geffroyite, 15a Manganoshadlunite, 15a Pentlandite, 15a Shadlunite, 15b Godlevskite, 15c Sugakiite; 20 Vozhminite
- o 02.BC With Rh, Pd, Pt, etc.: 05 Palladseite, 05 Miassite; 10 Oosterboschite; 15 Jagueite, 15 Chrisstanleyite; 20 Keithconnite, 25 Vasilite, 30 Telluropalladinitev, 35 Luberoite, 40 Oulankaite, 45 Telargpalite, 50 Temagamite, 55 Sopcheite, 60 Laflammeite, 65 Tischendorfite, 70 Kharaelakhite
- o 02.BD With Hg, Tl: 05 Imiterite, 10 Gortdrumite; 15 Balkanite, 15 Danielsite; 20 Donharrisite, 25 Carlinite; 30 Bukovite, 30 Thalcusite, 30 Murunskite; 35 Rohaite, 40 Chalcothallite, 45 Sabatierite, 50 Crookesite, 55 Brodtkorbite
- o 02.BE With Pb(Bi): 05 Betekhtinite, 10 Furutobeite; 15 Rhodplumsite, 15 Shandite; 20 Parkerite, 25 Schlemaite, 30 Pasavaite

- 02.C Metal Sulfides, M:S = 1:1 (and similar)
  - o 02.CA With Cu: 05a Covellite, 05b Klockmannite, 05c Spionkopite, 05d Yarrowite; 10 Nukundamite, 15 Calvertite
  - o 02.CB With Zn, Fe, Cu, Ag, Au, etc: 05a Rudashevskyite, 05a Hawleyite, 05a Coloradoite, 05a Metacinnabar, 05a Sphalerite, 05a Tiemannite, 05a Stilleite, 05b Sakuraiite, 05c Polhemusite; 07.0 Arsenosulvanite?; 10a Chalcopyrite, 10a Eskebornite, 10a Gallite, 10a Lenaite, 10a Roquesite, 10a Laforetite, 10b Haycockite, 10b Mooihoekite, 10b Putoranite, 10b Talnakhite; 15a Cernyite, 15a Hocartite, 15a Kuramite, 15a Pirquitasite, 15a Stannite, 15a Velikite, 15a Idaite, 15a Ferrokesterite, 15a Kesterite, 15b Mohite, 15c Stannoidite; 20 Chatkalite, 20 Mawsonite; 30 Colusite, 30 Germanite, 30 Germanocolusite, 30 Nekrasovite, 30 Stibiocolusite, 30 Maikainite, 30 Ovamboite; 35a Hemusite,

35a Kiddcreekite, 35a Renierite, 35a Polkovicite, 35a Morozeviczite, 35a Catamarcaite, 35a Vinciennite; 40 Lautite; 45 Cadmoselite, 45 Rambergite, 45 Greenockite, 45 Wurtzite; 55a Cubanite, 55b Isocubanite; 60 Picotpaulite, 60 Raguinite; 65 Argentopyrite, 65 Sternbergite; 70 Sulvanite, 75 Vulcanite, 80 Empressite, 85 Muthmannite

- o 02.CC With Ni, Fe, Co, PGE, etc.: 05 Zlatogorite, 05 Breithauptite, 05 Freboldite, 05 Langisite, 05 Nickeline, 05 Sederholmite, 05 Stumpflite, 05 Sudburyite, 05 Sobolevskite, 05 Achavalite, 05 Jaipurite*, 05 Hexatestibiopanickelite, 05 Kotulskite; 10 Smythite, 10 Pyrrhotite, 10 Troilite; 15 Cherepanovite, 15 Modderite, 15 Ruthenarsenite, 15 Westerveldite; 20 Makinenite, 20 Millerite; 25 Mackinawite, 30 Vavrinite; 35a Braggite, 35a Cooperite, 35a Vysotskite
- o 02.CD With Sn, Pb, Hg, etc.: 05 Herzenbergite, 05 Teallite; 10 Altaite, 10 Galena, 10 Clausthalite, 10 Alabandite, 10 Niningerite, 10 Oldhamite, 10 Keilite; 15a Cinnabar, 15b Hypercinnabar

- • 02.D Metal Sulfides, M:S = 3:4 and 2:3
  - o 02.DA M:S = 3:4: 05 Bornhardtite, 05 Florensovite, 05 Carrollite, 05 Fletcherite, 05 Daubreelite, 05 Greigite, 05 Linnaeite, 05 Kalininite, 05 Polydymite, 05 Violarite, 05 Tyrrellite, 05 Siegenite, 05 Trustedtite, 05 Cadmoindite, 05 Cuproiridsite, 05 Cuprorhodsite, 05 Dayingite*, 05 Ferrorhodsite, 05 Indite, 05 Malanite, 05 Xingzhongite; 10 Rhodostannite, 10 Toyohaite; 15 Wilkmanite, 15 Brezinaite, 15 Heideite; 20 Inaglyite, 20 Konderite; 25 Kingstonite
  - o 02.DB M:S = 2:3 and similar: 05 Heklaite, 05a Antimonselite, 05a Guanajuatite, 05a Bismuthinite, 05a Stibnite, 05a Metastibnite, 05b Paakkonenite; 10 Ottemannite, 10 Suredaite; 15 Bowieite, 15 Kashinite; 20 Montbrayite, 25 Edgarite, 30 Tarkianite, 35 Cameronite
  - o 02.DC Variable M:S: 05 Platynite?, 05a Hedleyite, 05b Nevskite, 05b Telluronevskite, 05b Ingodite, 05b Sulphotsumoite, 05b Tsumoite, 05c Kawazulite, 05c Paraguanajuatite, 05c Skippenite, 05c Tetradymite, 05c Tellurantimony, 05c Tellurobismuthite, 05d Laitakarite, 05d Ikunolite, 05d Joseite, 05d Joseite-B, 05d Pilsenite,

05e Vihorlatite, 05e Baksanite, 05e Protojoseite*, 05e Sztrokayite*

- 02.E Metal Sulfides, M:S £ 1:2
  - o 02.EA M:S = 1:2: 05 Sylvanite, 10 Calaverite; 15 Krennerite, 15 Kostovite; 20 Berndtite, 20 Merenskyite, 20 Melonite, 20 Kitkaite, 20 Moncheite, 20 Sudovikovite, 20 Shuangfengite; 25 Verbeekite; 30 Drysdallite, 30 Jordisite, 30 Molybdenite, 30 Tungstenite
  - o 02.EB M:S = 1:2, with Fe, Co, Ni, PGE, etc.: 05a Aurostibite, 05a Cattierite, 05a Hauerite, 05a Fukuchilite, 05a Erlichmanite, 05a Geversite, 05a Insizwaite, 05a Laurite, 05a Krutaite, 05a Pyrite, 05a Penroseite, 05a Sperrylite, 05a Vaesite, 05a Villamaninite, 05a Trogtalite, 05a Dzharkenite, 05a Gaotaiite, 05b Bambollaite; 10a Frohbergite, 10a Hastite?, 10a Ferroselite, 10a Kullerudite, 10a Mattagamite, 10a Marcasite, 10b Alloclasite, 10c Glaucodot, 10d Costibite, 10e Pararammelsbergite, 10e Paracostibite, 10f Oenite; 15a Clinosafflorite, 15a Anduoite, 15a Omeiite, 15a Lollingite, 15a Nisbite, 15a Rammelsbergite, 15a Safflorite, 15b Seinajokite; 20 Paxite, 20 Arsenopyrite, 20 Gudmundite, 20 Ruarsite, 20 Osarsite; 25 Krutovite, 25 Cobaltite, 25 Changchengite, 25 Hollingworthite, 25 Gersdorffite, 25 Irarsite, 25 Jolliffeite, 25 Padmaite, 25 Platarsite, 25 Ullmannite, 25 Tolovkite, 25 Willyamite, 25 Milotaite, 25 Kalungaite, 25 Maslovite, 25 Testibiopalladite, 25 Michenerite, 25 Mayingite; 30 Urvantsevite, 35 Rheniite
  - o 02.EC M:S = 1:>2: 05 Ferroskutterudite, 05 Kieftite, 05 Dienerite?, 05 Nickelskutterudite, 05 Skutterudite; 10 Patronite
- 02.F Sulfides of Arsenic, Alkalies; Sulfides with Halide, Oxide, Hydroxide, H_2O
  - o 02.FA With As, (Sb), S: 05 Duranusite, 10 Dimorphite, 15a Realgar, 15b Pararealgar, 20 Alacranite, 25 Uzonite; 30 Laphamite, 30 Orpiment; 35 Getchellite, 40 Wakabayashilite
  - o 02.FB With Alkalies (without Cl, etc.): 05 Cronusite, 05 Caswellsilverite, 05 Schollhornite; 10 Chvilevaite, 15 Orickite; 20 Rasvumite, 20 Pautovite; 25 Colimaite

- 02.FC With Cl, Br, I (halide-sulfides): 05 Djerfisherite, 05 Owensite, 05 Thalfenisite; 10 Bartonite, 10 Chlorbartonite; 15a Arzakite, 15a Corderoite, 15a Lavrentievite, 15b Kenhsuite, 15c Grechishchevite, 15d Radtkeite; 20a Capgaronnite, 20b Iltisite, 20c Perroudite; 25 Demicheleite-(Br), 25 Demicheleite-(Cl)
- 02.FD With O, OH, H_2O: 05 Kermesite, 10 Viaeneite, 20 Erdite, 25 Coyoteite; 30 Haapalaite, 30 Valleriite, 30 Yushkinite; 35 Tochilinite, 40 Wilhelmramsayite, 45 Vyalsovite, 50 Bazhenovite

- 02.X Unclassified Strunz Sulfides
  - 02.XX Unknown: 00 Horsfordite?, 00 Imgreite?, 00 Bravoite?, 00 Isochalcopyrite?, 00 Bayankhanite?, 00 Dzhezkazganite*, 00 Matraite?, 00 Iridisite*, 00 Prassoite, 00 Samaniite, 00 Horomanite, 00 Jeromite?, 00 Dilithium*, 00 Kurilite?

Sulfosalt Minerals

Sulfosalt minerals are those complex sulfide minerals with the general formula: $A_mB_nS_p$; where A represents a metal such as copper, lead, silver, iron and rarely mercury, zinc, vanadium; B usually represents semi-metal such as arsenic, antimony, bismuth and rarely germanium, or metals like tin and rarely vanadium; and S is sulfur or rarely selenium or/and tellurium. The Strunz classification includes the sulfosalts in a *sulfides and sulfosalts* superclass. A group which have a similar appearing formulas are the sulfarsenides (for example cobaltite (Co,Fe)AsS). In sulfarsenides the arsenic substitutes for sulfur whereas in the sulfosalts the arsenic substitutes for a metal cation.

There are about 200 known sulfosalt minerals. Examples include:

- A_3BX_3 type
  - Pyrargyrite Ag_3SbS_3
  - Proustite Ag_3AsS_3
  - Tetrahedrite $Cu_{12}Sb_4S_{13}$
  - Tennantite $Cu_{12}As_4S_{13}$
- A_3BX_4 type
  - Enargite Cu_3AsS_4
  - Sulvanite Cu_3VS_4

- Samsonite $Ag_4MnSb_2S_6$
- Geocronite $Pb_{14}(Sb,As)_6S_{23}$
- Gratonite $Pb_9As_4S_{15}$

- A_2BX_3 type
  - Bournonite $PbCuSbS_3$
  - Seligmannite $PbCuAsS_3$
  - Aikinite $PbCuBiS_3$
- ABX_2 type
  - Boulangerite $Pb_5Sb_4S_{11}$
  - Matildite $AgBiS_2$
  - Smithite $AgAsS_2$
  - Chalcostibite $CuSbS_2$
  - Emplectite $CuBiS_2$
  - Teallite $PbSnS_2$
- $A_2B_2X_5$ type
  - Ramdohrite $Ag_3Pb_6Sb_{11}S_{24}$
  - Jamesonite $Pb_4FeSb_6S_{14}$
  - Cosalite $Pb_2Bi_2S_5$
- $A_2B_3X_6$ type
  - Andorite $PbAgSb_3S_6$
  - Lindstromite $Pb_3Cu_3Bi_7S_{15}$
- AB_2X_4 type
  - Zinkenite $Pb_9Sb_{22}S_{42}$
  - Berthierite $FeSb_2S_4$
  - Cylindrite $Pb_3Sn_4FeSb_2S_{14}$.

Nickel˘Strunz Classification -02-Sulfosalts

IMA-CNMNC proposes a new hierarchical scheme (Mills et al., 2009). This list uses the Classification of Nickel–Strunz (10 ed, pending publication).

- Abbreviations:
  - "*"-discredited (IMA/CNMNC status).
  - "?"-questionable/doubtful (IMA/CNMNC status).
  - "REE"-Rare-earth element (Sc, Y, La, Ce, Pr, Nd, Pm, Sm, Eu, Gd, Tb, Dy, Ho, Er, Tm, Yb, Lu)

- o "PGE"-Platinum-group element (Ru, Rh, Pd, Os, Ir, Pt)
- o 03.C Aluminofluorides, 06 Borates, 08 Vanadates (04.H V Vanadates), 09 Silicates:
  - * Neso: insular
  - * Soro: grouping (from Greek óùñïæ sôros, heap, mound (especially of corn))
  - * Cyclo: ring
  - * Ino: chain
  - * Phyllo: sheet
  - * Tekto: three-dimensional framework
- Nickel–Strunz code scheme: NN.XY.##x
  - o NN: Nickel–Strunz mineral class number
  - o X: Nickel–Strunz mineral division letter
  - o Y: Nickel–Strunz mineral family letter
  - o ##x: Nickel–Strunz mineral/group number, x add-on letter.

Class: sulfosalts

- 02.G Sulfarsenites, sulfantimonites, sulfbismuthites
  - o 02.G: IMA2007-010
  - o 02.GA Neso-sulfarsenites, etc., without additional S: 05 Proustite, 05 Pyrargyrite; 10 Xanthoconite, 10 Pyrostilpnite; 15 Samsonite; 20 Wittichenite, 20 Skinnerite, 25 Malyshevite, 25 Lisiguangite, 25 Muckeite, 25 Lapieite; 30 Aktashite, 30 Nowackiite, 30 Gruzdevite; 35 Laffittite; 40 Stalderite, 40 Routhierite; 45 Erniggliite; 50 Seligmannite, 50 Soucekite, 50 Bournonite
  - o 02.GB Neso-sulfarsenites, etc: 05 Argentotennantite, 05 Giraudite, 05 Goldfieldite, 05 Freibergite, 05 Hakite, 05 Tennantite, 05 Tetrahedrite; 10 Selenostephanite, 10 Stephanite; 15 Cupropearceite, 15 Selenopolybasite, 15 Cupropolybasite, 15 Polybasite, 15 Pearceite, 15 Antimonpearceite, 15 Arsenpolybasite, 20 Galkhaite
  - o 02.GC Poly-sulfarsenites: 05 Hatchite, 05 Wallisite; 10 Sinnerite, 15 Watanabeite, 20 Simonite, 25 Q-ratite, 30 Smithite, 35 Trechmannite, 40a Aleksite, 40b Kochkarite, 40c Rucklidgeite, 40c Poubaite, 40d Saddlebackite, 40e Babkinite; 45 Tvalchrelidzeite, 50 Mutnovskite

- 02.H Sulfosalts of SnS Archetype
  - 02.HA With Cu, Ag, Fe (without Pb): 05 Emplectite, 05 Chalcostibite; 10 Miargyrite, 15 Livingstonite; 20 Berthierite, 20 Clerite, 20 Garavellite; 25 Baumstarkite, 25 Aramayoite
  - 02.HB With Cu, Ag, Hg, Fe, Sn and Pb: 05a Krupkaite, 05a Aikinite, 05a Hammarite, 05a Gladite, 05a Friedrichite, 05a Lindstromite, 05a Pekoite, 05a Paarite, 05a Emilite, 05a Salzburgite, 05b Meneghinite, 05c Jaskolskiite; 10a Kobellite, 10a Tintinaite, 10b Giessenite, 10b Izoklakeite, 10c Eclarite; 15 Jamesonite, 15 Benavidesite; 20a Nagyagite, 20b Buckhornite, 20c Museumite, 20d Berryite, 20e Watkinsonite
  - 02.HC With only Pb: 05a Sartorite, 05a Twinnite, 05a Guettardite, 05b Baumhauerite, 05b Baumhauerite-2a, 05c Liveingite, 05d Dufrenoysite, 05d Veenite, 05d Rathite, 05e Chabourneite, 05f Pierrotite, 05f Parapierrotite, 05g Marumoite; 10a Fuloppite, 10b Bismutoplagionite*, 10b Plagionite, 10c Heteromorphite, 10d Semseyite, 10d Rayite; 15 Boulangerite, 15 Falkmanite, 15 Plumosite*; 20 Robinsonite, 25 Moeloite, 30 Dadsonite, 35 Zoubekite, 35 Owyheeite
  - 02.HD With Tl: 05 Lorandite, 05 Weissbergite; 15 Christite, 20 Jankovicite, 25 Rebulite, 30 Imhofite, 35 Edenharterite, 40 Jentschite, 45 Hutchinsonite, 50 Bernardite, 55 Sicherite, 60 Gabrielite
  - 02.HE With alkalies, H_2O: 05 Gerstleyite
  - 02.HF With SnS and PbS archetype structure units: 20 Vrbaite; 25a Abramovite, 25a Levyclaudite, 25a Cylindrite, 25b Coiraite, 25b Incaite, 25b Potosiite, 25b Franckeite; 30 Lengenbachite
- 02.J Sulfosalts of PbS Archetype
  - 02.JA Galena derivatives with little or no Pb: 05a IMA2005-036, 05a IMA2008-058, 05a Cupropavonite, 05a Pavonite, 05b Grumiplucite, 05c Kudriavite, 05d Cupromakovickyite, 05d Makovickyite, 05e Benjaminite, 05f Mummeite, 05g Borodaevite, 05h Mozgovaite; 10a Cuprobismutite, 10b Kupcikite, 10c Hodrushite, 10d Pizgrischite, 10e Paderaite;

15 Cuboargyrite, 15 Schapbachite; 20 Bohdanowiczite, 20 Matildite, 20 Volynskite

- 02.JB Galena derivatives, with Pb: 05 Diaphorite, 10 Cosalite; 15 Marrite, 15 Freieslebenite; 20 Cannizzarite, 20 Wittite; 25a Junoite, 25b Felbertalite, 25c Nordstromite, 25d Proudite, 25g Nuffieldite, 25i IMA2008-053, 25i Neyite, 25j Rouxelite; 30a Jordanite, 30a Geocronite, 30b Kirkiite, 30c Tsugaruite; 35a Zinkenite, 35b Scainiite, 35c Pillaite, 35d Pellouxite; 40a Bursaite?, 40a Gustavite, 40a Lillianite, 40a Xilingolite, 40a Treasurite, 40a Vikingite, 40a Fizelyite, 40a Andorite, 40a Roshchinite, 40a Uchucchacuaite, 40a Ramdohrite, 40b Aschamalmite, 40b Eskimoite, 40b Heyrovskyite, 40c Ourayite, 40d Schirmerite, 40e Ustarasite; 45 Angelaite, 45 Galenobismutite, 45 Weibullite; 55 Gratonite, 60 Marrucciite, 65 Vurroite
- 02.JC Galena derivatives, with Tl: 05 Ellisite, 10 Gillulyite

- 02.K Sulfarsenates, Sulfantimonates
  - 02.KA Sulfarsenates with $(As,Sb)S_4$ tetrahedra: 05 Enargite, 05 Stibioenargite*, 05 Petrukite; 10 Briartite, 10 Famatinite, 10 Luzonite, 10 Permingeatite, 10 Barquillite; 15 Fangite
  - 02.KB Sulfarsenates with additional S: 05 Billingsleyite
- 02.L Unclassified Sulfosalts
  - 02.LA Without essential Pb: 10 Dervillite, 15 Daomanite*, 20 Vaughanite, 25 Criddleite, 30 Fettelite, 35 Chameanite, 40 Arcubisite, 45 Mgriite, 50 Benleonardite, 55 Tsnigriite, 60 Borovskite, 65 Jonassonite
  - 02.LB With essential Pb: 05 Miharaite, 20 Ardaite, 30 Madocite, 35 Larosite; 40 Petrovicite, 40 Mazzettiite; 45 Crerarite, 50 Launayite, 55 Playfairite, 60 Sorbyite, 65 Sterryite
- 02.M
  - 02.MA Oxysulfosalts of Alkalies and Alkali Earths: 05 Ottensite, 05 Cetineite; 10 Sarabauite
- 02.X Unclassified Strunz Sulfosalts
  - 02.XX Unknown: 00 Tazieffite, 00 Horobetsuite*, 00 Kitaibelite?, 00 Parajamesonite?, 00 Sakharovaite?, 00 Volfsonite*

Halide Minerals

The halide mineral class include those minerals with a dominant halide anion (F^-, Cl^-, Br^- and I^-). Complex halide minerals may also have polyatomic anions in addition to or that include halides.

Fluorite structure

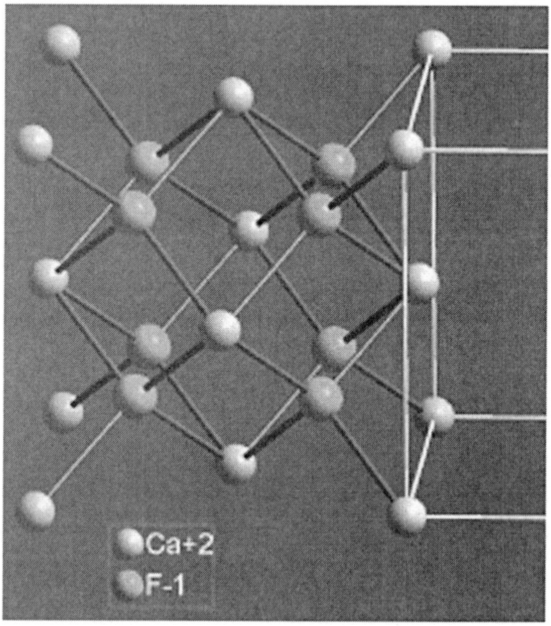

Figure 1 : Fluorite structure

Examples include the following:

- Halite NaCl
- Sylvite KCl
- Chlorargyrite AgCl and bromargyrite AgBr
- Fluorite CaF_2
- Atacamite $Cu_2Cl(OH)_3$
- Bischofite ($MgCl_2 \cdot 6H_2O$)
- Carnallite $KMgCl_3 \cdot 6H_2O$
- Cryolite Na_3AlF_6
- Cryptohalite (α)$(NH_4)_2SiF_6$
- Bararite (β)$(NH_4)_2SiF_6$

Commercially Significant Halide minerals

Two commercially important halide minerals are halite and

fluorite. The former is a major source of sodium chloride, in parallel with sodium chloride extracted from sea water or brine wells. Fluorite is a major source of hydrogen fluoride, complementing the supply obtained as a byproduct of the production of fertilizer. Carnallite and bischofite are important sources of magnesium. Natural cryolite was historically required for the production of aluminium, however, currently most cryolite used is produced synthetically.

Many of the halide minerals occur in marine evaporite deposits.

Nickel˘Strunz Classification-03-Halides

IMA-CNMNC proposes a new hierarchical scheme. This list uses the Classification of Nickel–Strunz (10 ed, pending publication).

- Abbreviations:
  - o "*"-discredited (IMA/CNMNC status).
  - o "?"-questionable/doubtful (IMA/CNMNC status).
  - o "REE"-Rare-earth element (Sc, Y, La, Ce, Pr, Nd, Pm, Sm, Eu, Gd, Tb, Dy, Ho, Er, Tm, Yb, Lu)
  - o "PGE"-Platinum-group element (Ru, Rh, Pd, Os, Ir, Pt)
  - o 03.C Aluminofluorides, 06 Borates, 08 Vanadates (04.H V Vanadates), 09 Silicates:
    - * Neso: insular
    - * Soro: grouping (from Greek óùñïæ sôros, heap, mound (especially of corn))
    - * Cyclo: ring
    - * Ino: chain
    - * Phyllo: sheet
    - * Tekto: three-dimensional framework
- Nickel–Strunz code scheme: NN.XY.##x
  - o NN: Nickel–Strunz mineral class number
  - o X: Nickel–Strunz mineral division letter
  - o Y: Nickel–Strunz mineral family letter
  - o ##x: Nickel–Strunz mineral/group number, x add-on letter.

Class: Halides

- 03.A Simple Halides, Without H_2O
  - o 03.AA M:X = 1:1, 2:3, 3:5, etc.: Panichiite; 05 Nantokite, 05 Marshite, 05 Miersite; 10 Iodargyrite, 10 Tocornalite;

15 Bromargyrite, 15 Embolite*, 15 Chlorargyrite; 20 Carobbiite, 20 Griceite, 20 Halite, 20 Sylvite, 20 Villiaumite; 25 Salammoniac, 25 Lafossaite; 30 Calomel, 30 Kuzminite, 30 Moschelite; 35 Neighborite; 40 Chlorocalcite, 45 Kolarite, 50 Radhakrishnaite; 55 Hephaistosite, 55 Challacolloite

- o 03.AB M:X = 1:2: 05 Tolbachite, 10 Coccinite, 15 Sellaite; 20 Chloromagnesite*, 20 Lawrencite, 20 Scacchite; 25 Frankdicksonite, 25 Fluorite; 30 Tveitite-(Y); 35 Gagarinite-(Y); 35 Zajacite-(Ce)
- o 03.AC M:X = 1:3: 05 Zharchikhite, 10 Molysite; 15 Fluocerite-(Ce), 15 Fluocerite-(La), 20 Gananite

- 03.B Simple Halides, with H_2O
  - o 03.BA M:X = 1:1 and 2:3: 05 Hydrohalite, 10 Carnallite
  - o 03.BB M:X = 1:2: 05 Eriochalcite, 10 Rokuhnite, 15 Bischofite, 20 Nickelbischofite, 25 Sinjarite, 30 Antarcticite, 35 Tachyhydrite
  - o 03.BC M:X = 1:3: 05 Chloraluminite
  - o 03.BD Simple Halides with H_2O and additional OH: 05 Cadwaladerite, 10 Lesukite, 15 Korshunovskite, 20 Nepskoeite, 25 Koenenite
- 03.C Complex Halides
  - o 03.C: Steropesite, IMA2008-032, IMA2008-039
  - o 03.CA Borofluorides: 05 Ferruccite; 10 Avogadrite, 10 Barberiite
  - o 03.CB Neso-aluminofluorides: 05 Cryolithionite; 15 Cryolite, 15 Elpasolite, 15 Simmonsite; 20 Colquiriite, 25 Weberite, 30 Karasugite, 35 Usovite; 40 Pachnolite, 40 Thomsenolite; 45 Carlhintzeite, 50 Yaroslavite
  - o 03.CC Soro-aluminofluorides: 05 Gearksutite; 10 Acuminite, 10 Tikhonenkovite; 15 Artroeite; 20 Calcjarlite, 20 Jarlite, 20 Jorgensenite
  - o 03.CD Ino-aluminofluorides: 05 Rosenbergite, 10 Prosopite
  - o 03.CE Phyllo-aluminofluorides: 05 Chiolite
  - o 03.CF Tekto-aluminofluorides: 05 Ralstonite, 10 Boldyrevite?, 15 Bogvadite
  - o 03.CG Aluminofluorides with CO_3, SO_4, PO_4: 05 Stenonite; 10 Chukhrovite-(Nd), 10 Chukhrovite-(Ce), 10 Chukhrovite-

(Y), 10 Meniaylovite; 15 Creedite, 20 Boggildite, 25 Thermessaite

- o 03.CH: 05 Malladrite, 10 Bararite; 15 Cryptohalite, 15 Hieratite; 20 Demartinite, 25 Knasibfite
- o 03.CJ With MX_6 complexes; M = Fe, Mn, Cu: 05 Chlormanganokalite, 05 Rinneite; 10 Erythrosiderite, 10 Kremersite; 15 Mitscherlichite, 20 Douglasite, 30 Zirklerite

- 03.D Oxyhalides, Hydroxyhalides and Related Double Halides
  - o 03.DA With Cu, etc., without Pb: 05 Melanothallite; 10a Atacamite, 10a Kempite, 10a Hibbingite, 10b Botallackite, 10b Clinoatacamite, 10b Belloite, 10c Gillardite, 10c Kapellasite, 10c Haydeeite, 10c Paratacamite, 10c Herbertsmithite; 15 Claringbullite, 20 Simonkolleite; 25 Buttgenbachite, 25 Connellite; 30 Abhurite, 35 Ponomarevite; 40 Calumetite, 40 Anthonyite; 45 Khaidarkanite, 50 Bobkingite, 55 Avdoninite, 60 Droninoite
  - o 03.DB With Pb, Cu, etc.: 05 Diaboleite, 10 Pseudoboleite, 15 Boleite, 20 Cumengite, 25 Bideauxite, 30 Chloroxiphite, 35 Hematophanite; 40 Asisite, 40 Parkinsonite; 45 Murdochite, 50 Yedlinite
  - o 03.DC With Pb (As, Sb, Bi), without Cu: 05 Laurionite, 05 Paralaurionite; 10 Fiedlerite, 15 Penfieldite, 20 Laurelite; 25 Zhangpeishanite, 25 Matlockite, 25 Rorisite, 25 Daubreeite, 25 Bismoclite, 25 Zavaritskite; 30 Nadorite, 30 Perite; 35 Aravaipaite, 37 Calcioaravaipaite, 40 Thorikosite, 45 Mereheadite, 50 Blixite, 55 Pinalite, 60 Symesite; 65 Ecdemite, 65 Heliophyllite; 70 Mendipite, 75 Damaraite, 80 Onoratoite, 85 Cotunnite, 90 Pseudocotunnite, 95 Barstowite
  - o 03.DD With Hg: 05 Eglestonite, 05 Kadyrelite; 10 Poyarkovite, 15 Hanawaltite, 20 Terlinguaite, 25 Pinchite; 30 Mosesite, 30 Gianellaite; 35 Kleinite, 40 Tedhadleyite, 45 Vasilyevite, 50 Aurivilliusite, 55 Terlinguacreekite, 60 Kelyanite, 65 Comancheite
  - o 03.DE With Rare-Earth Elements: 05 Haleniusite-(La)
- 03.X Unclassified Strunz Halogenides
  - o 03.XX Unknown: 00 Hydrophilite?, 00 Hydromolysite?, 00 Yttrocerite*, 00 Lorettoite?, 00 IMA2009-014, 00 IMA2009-015

Oxide Minerals

The oxide mineral class includes those minerals in which the oxide anion (O^{2-}) is bonded to one or more metal ions. The hydroxide bearing minerals are typically included in the oxide class. The minerals with complex anion groups such as the silicates, sulfates, carbonates and phosphates are classed separately.

Simple oxides:

- X_2O and XO
    - o Cuprite Cu_2O
    - o Ice H_2O
    - o Periclase group
        - * Periclase MgO
        - * Manganosite MnO
    - o Zincite group
        - * Zincite ZnO
        - * Bromellite BeO
        - * Tenorite CuO
        - * Litharge PbO
- X_2O_3
    - o Hematite group
        - * Corundum Al_2O_3
        - * Hematite Fe_2O_3
        - * Ilmenite $FeTiO_3$
- XO_2
    - o Rutile group
        - * Rutile TiO_2
        - * Pyrolusite MnO_2
        - * Cassiterite SnO_2
        - o Baddeleyite ZrO_2
        - o Uraninite UO_2
        - o Thorianite ThO_2
- XY_2O_4
    - o Spinel group
        - * Spinel $MgAl_2O_4$

    * Gahnite $ZnAl_2O_4$
    * Magnetite Fe_3O_4 ($Fe^{2+}Fe^{3+}{}_2O_4$)
    * Franklinite $(Zn,Fe,Mn)(Fe,Mn)_2O_4$
    * Chromite $FeCr_2O_4$
  - o Chrysoberyl $BeAl_2O_4$
  - o Columbite$(Fe,Mn)(Nb,Ta)_2O_6$

Hyroxide subgroup:

- Brucite $Mg(OH)_2$
- Manganite $MnO(OH)$
- Romanechite $BaMn^{2+}Mn^{4+}{}_8O_{16}(OH)_4$
- Goethite group
  - o Diaspore $\alpha AlO(OH)$
  - o Goethite $\alpha FeO(OH)$.

Nickel˘strunz Classification-04-Oxides

IMA-CNMNC proposes a new hierarchical scheme. This list uses it to modify the Classification of Nickel–Strunz (10 ed, pending publication).

- Abbreviations:
  - o "*"-discredited (IMA/CNMNC status).
  - o "?"-questionable/doubtful (IMA/CNMNC status).
  - o "REE"-Rare-earth element (Sc, Y, La, Ce, Pr, Nd, Pm, Sm, Eu, Gd, Tb, Dy, Ho, Er, Tm, Yb, Lu)
  - o "PGE"-Platinum-group element (Ru, Rh, Pd, Os, Ir, Pt)
  - o 03.C Aluminofluorides, 06 Borates, 08 Vanadates (04.H V Vanadates), 09 Silicates:
    * Neso: insular
    * Soro: grouping
    * Cyclo: ring
    * Ino: chain
    * Phyllo: sheet
    * Tekto: three-dimensional framework
- Nickel–Strunz code scheme: NN.XY.##x
  - o NN: Nickel–Strunz mineral class number
  - o X: Nickel–Strunz mineral division letter

- o Y: Nickel–Strunz mineral family letter
- o ##x: Nickel–Strunz mineral/group number, x add-on letter.

Class: Oxides

- 04.A Metal:Oxygen = 2.1 and 1:1
  - o 04.AA Cation:Anion (M:O) = 2:1 (and 1.8:1): 05 Ice, 10 Cuprite, 15 Paramelaconite
  - o 04.AB M:O = 1:1 (and up to 1:1.25); with small to medium-sized cations only: 05 Crednerite, 10 Tenorite; 15 Delafossite, 15 Mcconnellite; 20 Bromellite, 20 Zincite; 25 Lime, 25 Bunsenite, 25 Monteponite, 25 Manganosite, 25 Periclase, 25 Wustite
  - o 04.AC M:O = 1:1 (and up to 1:1.25); with large cations (± smaller ones): 05 Swedenborgite; 10 Brownmillerite, 10 Srebrodolskite; 15 Montroydite, 20 Litharge, 20 Romarchite, 25 Massicot
- 04.B Metal:Oxygen = 3:4 and similar
  - o 04.BA With small and medium-sized cations: 05 Chrysoberyl, 10 Manganostibite
  - o 04.BB With only medium-sized cations: 05 Filipstadite, 05 Donathite?, 05 Gahnite, 05 Galaxite, 05 Hercynite, 05 Spinel, 05 Cochromite, 05 Chromite, 05 Magnesiochromite, 05 Manganochromite, 05 Nichromite, 05 Zincochromite, 05 Magnetite, 05 Cuprospinel, 05 Franklinite, 05 Jacobsite, 05 Magnesioferrite, 05 Trevorite, 05 Brunogeierite, 05 Coulsonite, 05 Magnesiocoulsonite, 05 Qandilite, 05 Ulvospinel, 05 Vuorelainenite; 10 Hydrohetaerolite, 10 Hausmannite, 10 Iwakiite, 10 Hetaerolite; 15 Maghemite, 20 Tegengrenite, 25 Xieite
  - o 04.BC With medium-sized and large cations: 05 Marokite, 10 Dmitryivanovite
  - o 04.BD With only large cations: 05 Minium
- 04.C Metal:Oxygen = 2:3, 3:5, and Similar
  - o 04.CB With medium-sized cations: 05 Tistarite, 05 Auroantimonate*, 05 Brizziite-VII, 05 Brizziite-III, 05 Corundum, 05 Eskolaite, 05 Hematite, 05 Karelianite, 05 Geikielite, 05 Ecandrewsite, 05 Ilmenite, 05 Pyrophanite, 05 Melanostibite, 05 Romanite*; 10 Bixbyite, 10 Avicennite;

15 Armalcolite, 15 Mongshanite*, 15 Pseudobrookite; 20 Magnesiohogbomite-6N6S, 20 Magnesiohogbomite-2N3S, 20 Magnesiohogbomite-2N2S, 20 Zincohogbomite-2N2S, 20 Ferrohogbomite-2N2S; 25 Pseudorutile, 25 Ilmenorutile; 30 Oxyvanite, 30 Berdesinskiite; 35 Olkhonskite, 35 Schreyerite; 40 Kamiokite, 40 Nolanite, 40 Rinmanite; 45 Stibioclaudetite, 45 Claudetite; 50 Arsenolite, 50 Senarmontite; 55 Valentinite, 60 Bismite, 65 Sphaerobismoite, 70 Sillenite, 75 Kyzylkumite

- o 04.CC With large and medium-sized cations: 05 Chrombismite, 10 Freudenbergite, 15 Grossite, 20 Mayenite, 25 Yafsoanite; 30 Barioperovskite, 30 Lakargiite, 30 Natroniobite, 30 Latrappite, 30 Lueshite, 30 Perovskite; 35 Macedonite, 35 Isolueshite, 35 Loparite-(Ce), 35 Tausonite; 40 Crichtonite, 40 Dessauite, 40 Davidite-(Ce), 40 Davidite-(La), 40 Mathiasite, 40 Lindsleyite, 40 Landauite, 40 Loveringite, 40 Loveringite, 40 Cleusonite, 40 Gramaccioliite-(Y); 45 Hawthorneite, 45 Magnetoplumbite, 45 Haggertyite, 45 Batiferrite, 45 Hibonite, 45 Nezilovite, 45 Yimengite, 45 Diaoyudaoite, 45 Lindqvistite, 45 Plumboferrite; 50 Jeppeite, 55 Zenzenite, 60 Mengxianminite*

- • 04.D Metal:Oxygen = 1:2 and similar
  - o 04.DA With small cations
    - * (moved to-09-Subclass: tektosilicates)
  - o 04.DB With medium-sized cations; chains of edge-sharing octahedra: 05 Tripuhyite, 05 Tugarinovite, 05 Varlamoffite*, 05 Argutite, 05 Cassiterite, 05 Rutile, 05 Pyrolusite, 05 Plattnerite, 05 Squawcreekite?; 10 Bystromite, 10 Ordonezite, 10 Tapiolite-(Fe), 10 Tapiolite-(Mn), 10 Tapiolite*, 15a Paramontroseite, 15a Ramsdellite, 15b Akhtenskite, 15c Nsutite; 20 Scrutinyite; 25 Ixiolite, 25 Ishikawaite, 25 Srilankite, 25 Samarskite-(Y), 25 Samarskite-(Yb), 25 Yttrocolumbite-(Y); 30 Heftetjernite, 30 Wolframoixiolite*, 30 Krasnoselskite*, 30 Ferberite, 30 Hubnerite, 30 Sanmartinite, 30 Wolframite*; 35 Tantalite-(Mg), 35 Tantalite-(Fe), 35 Tantalite-(Mn), 35 Columbite-(Mg), 35 Columbite-(Fe), 35 Columbite-(Mn), 35 Qitianlingite; 40 Ferrowodginite, 40 Lithiotantite, 40

Lithiowodginite, 40 Tantalowodginite*, 40 Titanowodginite, 40 Wodginite, 40 Ferrotitanowodginite; 45 Tivanite, 50 Carmichaelite, 55 Alumotantite, 60 Biehlite

- o 04.DC With medium-sized cations; sheets of edge-sharing octahedra: 05 Bahianite, 10 Simpsonite
- o 04.DD With medium-sized cations; frameworks of edge-sharing octahedra: 05 Anatase, 10 Brookite
- o 04.DE With medium-sized cations; with various polyhedra: 05 Downeyite, 10 Koragoite; 15 Koechlinite, 15 Russellite, 15 Tungstibite; 20 Tellurite, 25 Paratellurite; 30 Cervantite, 30 Bismutotantalite, 30 Bismutocolumbite, 30 Clinocervantite, 30 Stibiocolumbite, 30 Stibiotantalite; 35 IMA2007-058, 35 Baddeleyite
- o 04.DF With large (± medium-sized) cations; dimers and trimers of edge-sharing octahedra: 05 Nioboaeschynite-(Y), 05 Aeschynite-(Ce), 05 Aeschynite-(Nd), 05 Aeschynite-(Y), 05 Nioboaeschynite-(Ce), 05 Nioboaeschynite-(Nd), 05 Tantalaeschynite-(Y), 05 Rynersonite, 05 Vigezzite, 10 Changbaiite, 15 Murataite
- o 04.DG With large (± medium-sized) cations; chains of edge-sharing octahedra: 05 Euxenite-(Y), 05 Loranskite-(Y), 05 Polycrase-(Y), 05 Uranopolycrase, 05 Fersmite, 05 Kobeite-(Y), 05 Tanteuxenite-(Y), 05 Yttrocrasite-(Y); 10 Fergusonite-beta-(Nd), 10 Fergusonite-beta-(Y), 10 Fergusonite-beta-(Ce), 10 Yttrotantalite-(Y); 15 Foordite, 15 Thoreaulite; 20 Raspite
- o 04.DH With large (± medium-sized) cations; sheets of edge-sharing octahedra:
  - * IMA/CNMNC revised the Pyrochlore supergroup 2010 (04.DH.15 and 04.DH.20)
  - * 05 Brannerite, 05 Orthobrannerite, 05 Thorutite; 10 Kassite, 10 Lucasite-(Ce)
  - * Pyrochlore group: Fluorcalciopyrochlore, Fluorkenopyrochlore, Fluornatropyrochlore, Fluorstrontiopyrochlore, Hydropyrochlore, Hydroxycalciopyrochlore, Kenoplumbopyrochlore, Oxycalciopyrochlore, Oxynatropyrochlore, Oxyplumbopyrochlore, Oxyyttropyrochlore-(Y)
  - * Microlite group: Fluorcalciomicrolite, Fluornatromicrolite, Hydrokenomicrolite, Hydromicrolite,

Hydroxykenomicrolite, Kenoplumbomicrolite, Oxycalciomicrolite, Oxystannomicrolite, Oxystibiomicrolite

- * Romeite group: Cuproromeite, Fluorcalcioromeite, Fluornatroromeite, Hydroxycalcioromeite, Oxycalcioromeite, Oxyplumboromeite, Stibiconite
- * Betafite group: Calciobetafite, Oxyuranobetafite
- * Elsmoreite group: Hydrokenoelsmoreite
- * 25 Rosiaite; 30 Zirconolite-3O, 30 Zirconolite-3T, 30 Zirconolite-2M, 30 Zirconolite; 35 Liandratite, 35 Petscheckite; 40 Ingersonite, 45 Pittongite
- * Discredited minerals 04.DH.15: Bariomicrolite (of Hogarth 1977), Bariopyrochlore (of Hogarth 1977), Betafite (of Hogarth 1977), Bismutomicrolite (of Hogarth 1977), Ceriopyrochlore (of Hogarth 1977), Jixianite, Natrobistantite, Plumbomicrolite (of Hogarth 1977), Plumbobetafite (of Hogarth 1977), Stannomicrolite (of Hogarth 1977), Stibiobetafite (of Èerný et al.), Yttrobetafite (of Hogarth 1977), Yttropyrochlore (of Hogarth 1977), Bismutopyrochlore (of Chukanov et al.) and Bismutostibiconite 04.DH.20

- o 04.DJ With large (± medium-sized) cations; polyhedral frameworks: 05 Calciotantite, 05 Irtyshite, 05 Natrotantite
- o 04.DK With large (± medium-sized) cations; tunnel structures: 05 Ankangite, 05 Coronadite, 05 Hollandite, 05 Manjiroite, 05 Mannardite, 05 Redledgeite, 05 Priderite, 05 Henrymeyerite, 05 Akaganeite, 10 Cryptomelane, 10 Romanechite, 10 Strontiomelane, 10 Todorokite
- o 04.DL With large (± medium-sized) cations; fluorite-type structures: 05 Cerianite-(Ce), 05 Zirkelite, 05 Thorianite, 05 Uraninite; 10 Calzirtite, 10 Hiarneite, 10 Tazheranite
- o 04.DM With large (± medium-sized) cations; unclassified: 05 Sosedkoite, 05 Rankamaite; 15 Cesplumtantite, 20 Eyselite, 25 Kuranakhite

- • 04.E Metal:Oxygen = < 1:2
  - o 04.E: IMA2008-040
  - o 04.EA Oxides with metal : oxygen < 1:2 (M_2O_5, MO_3): 05 Tantite, 10 Krasnogorite*, 10 Molybdite

- 04.X Unclassified Strunz Oxides
  - 04.XX Unknown: 00 Allendeite, 00 Ashanite?, 00 Hongquiite*, 00 Psilomelane?, 00 Uhligite?, 00 Clinobirnessite*, 00 Kleberite*, 00 Chubutite*, 00 Struverite?, 00 IMA2000-016, 00 IMA2000-026

Class: hydroxides

- 04.F Hydroxides (without V or U)
  - 04.FA Hydroxides with OH, without H_2O; corner-sharing tetrahedra: 05a Behoite, 05b Clinobehoite; 10 Sweetite, 10 Wulfingite, 10 Ashoverite
  - 04.FB Hydroxides with OH, without H_2O; insular octahedra: 05 Shakhovite; 10 Cualstibite, 10 Zincalstibite
  - 04.FC Hydroxides with OH, without H_2O; corner-sharing octahedra: 05 Dzhalindite, 05 Sohngeite, 05 Bernalite; 10 Burtite, 10 Mushistonite, 10 Natanite, 10 Vismirnovite, 10 Schoenfliesite, 10 Wickmanite; 15 Jeanbandyite, 15 Mopungite, 15 Stottite; 15 Tetrawickmanite; 20 Ferronigerite-6N6S, 20 Ferronigerite-2N1S, 20 Magnesionigerite-6N6S, 20 Magnesionigerite-2N1S; 25 Magnesiotaaffeite-6N3S, 25 Magnesiotaaffeite-2N2S, 25 Ferrotaaffeite-6N3S
  - 04.FD Hydroxides with OH, without H_2O; chains of edge-sharing octahedra: 05 Spertiniite; 10 Bracewellite, 10 Diaspore, 10 Guyanaite, 10 Groutite, 10 Goethite, 10 Montroseite, 10 Tsumgallite; 15 Manganitev; 20 Cerotungstite-(Ce), 20 Yttrotungstite-(Y), 20 Yttrotungstite-(Ce); 25 Frankhawthorneite; 30 Khinite, 30 Parakhinite
  - 04.FE Hydroxides with OH, without H_2O; sheets of edge-sharing octahedra: 05 Amakinite, 05 Brucite, 05 Portlandite, 05 Pyrochroite, 05 Theophrastite, 05 Fougerite; 10 Bayerite, 10 Doyleite, 10 Gibbsite, 10 Nordstrandite; 15 Boehmite, 15 Lepidocrocite; 20 Grimaldiite, 20 Heterogenite-2H, 20 Heterogenite-3R; 25 Feitknechtite, 25 Lithiophorite; 30 Quenselite, 35 Ferrihydrite; 40 Feroxyhyte, 40 Vernadite; 45 Quetzalcoatlite
  - 04.FF Hydroxides with OH, without H_2O; various polyhedra: 05 Hydroromarchite
  - 04.FG Hydroxides with OH, without H_2O; unclassified: 05 Janggunite, 10 Cesarolite, 15 Kimrobinsonite

- 04.FH Hydroxides with H_2O ± (OH); insular octahedra: 05 Bottinoite, 05 Brandholzite
- 04.FJ Hydroxides with H_2O ± (OH); corner-sharing octahedra: 05 Sidwillite, 05 Meymacite; 10 Tungstite; 15 Ilsemannite, 15 Hydrotungstite; 20 Parabariomicrolite
- 04.FK Hydroxides with H_2O ± (OH); chains of edge-sharing octahedra: 05 Bamfordite
- 04.FL Hydroxides with H_2O ± (OH); sheets of edge-sharing octahedra: 05 Meixnerite, 05 Jamborite, 05 Iowaite, 05 Woodallite, 05 Akdalaite, 05 Muskoxite; 10 Hydrocalumite, 15 Kuzelite; 20 Aurorite, 20 Chalcophanite, 20 Ernienickelite, 20 Jianshuiite; 25 Woodruffite, 30 Asbolane; 40 Takanelite, 40 Rancieite; 45 Birnessite, 55 Cianciulliite, 60 Jensenite, 65 Leisingite, 75 Cafetite, 80 Mourite, 85 Deloryite
- 04.FM Hydroxides with H_2O ± (OH); Unclassified: 15 Franconite, 15 Hochelagaite, 15 Ternovite; 25 Belyankinite, 25 Gerasimovskite, 25 Manganbelyankinite; 30 Silhydrite, 35 Cuzticite, 40 Cyanophyllite
- 04.FN: 05 Menezesite

- 04.G Uranyl Hydroxides
  - 04.GA Without additional cations: IMA2008-022; 05 Metaschoepite, 05 Paraschoepite, 05 Schoepite; 10 Ianthinite; 15 Metastudtite, 15 Studtite
  - 04.GB With additional cations (K, Ca, Ba, Pb, etc.); with mainly $UO_2(O,OH)_5$ pentagonal polyhedra: 05 Compreignacite, 05 Agrinierite, 05 Rameauite; 10 Billietite, 10 Becquerelite, 10 Protasite; 15 Richetite; 20 Calciouranoite, 20 Bauranoite, 20 Metacalciouranoite; 25 Fourmarierite, 30 Wolsendorfite, 35 Masuyite; 40 Metavandendriesscheite, 40 Vandendriesscheite; 45 Vandenbrandeite, 50 Sayrite, 55 Curite, 60 Iriginite, 65 Uranosphaerite, 70 Holfertite
  - 04.GC With additional cations; with $UO_2(O,OH)_6$ hexagonal polyhedra: 05 Clarkeite, 10 Umohoite, 15 Spriggite
- 04.H V Vanadates
  - (moved to-08-Class: vanadates)
- 04.X Unclassified Strunz Oxides (Hydroxides)
  - 04.XX Unknown: 00 Ungursaite*, 00 Scheteligite?

Arsenite Minerals

Arsenite minerals are very rare oxygen-bearing arsenic minerals. Classical world localities where such minerals occur include the complex skarn manganese deposit at Långban (Sweden) and the polymetallic Tsumeb deposit (Namibia). The most often reported arsenite anion in minerals is the AsO_3^{3-}anion, present for example in reinerite $Zn_3(AsO_3)_2$. Unique diarsenite anions occur i.e. in leiteite $Zn[As_2O_4]$ and paulmooreite $Pb[As_2O_5]$. More complex arsenites include schneiderhöhnite $Fe^{2+}Fe^{3+}_3[As_5O_{13}]$ and ludlockite $PbFe^{3+}_4As_{10}O_{22}$.

Nickel˜Strunz Classification-04- Oxides

IMA-CNMNC proposes a new hierarchical scheme (Mills et al., 2009). This list uses it to modify the Classification of Nickel–Strunz (10 ed, pending publication).

- Abbreviations:
  - "*"-IMA/CNMNC status: discredited.
  - "?"-IMA/CNMNC status: questionable/doubtful.
  - "REE"-Rare-earth element (Sc, Y, La, Ce, Pr, Nd, Pm, Sm, Eu, Gd, Tb, Dy, Ho, Er, Tm, Yb, Lu)
  - "PGE"-Platinum-group element (Ru, Rh, Pd, Os, Ir, Pt)
  - 03.C Aluminofluorides, 06 Borates, 08 Vanadates (04.H V Vanadates), 09 Silicates:
    - Neso: insular
    - Soro: grouping
    - Cyclo: ring
    - Ino: chain
    - Phyllo: sheet
    - Tekto: three-dimensional framework
- Nickel–Strunz code scheme: NN.XY.##x
  - NN: Nickel–Strunz mineral class number
  - X: Nickel–Strunz mineral division letter
  - Y: Nickel–Strunz mineral family letter
  - ##x: Nickel–Strunz mineral/group number, x add-on letter.

Class: Arsenites

- 04.J Arsenites, Antimonites, Bismuthites, Sulfites
  - 04.JA Arsenites, antimonites, bismuthites; without additional anions, without H_2O: 05 Leiteite, 10 Reinerite,

15 Karibibite; 20 Schafarzikite, 20 Trippkeite, 20 Kusachiite; 25 Apuanite, 30 Versiliaite, 35 Schneiderhohnite, 40 Zimbabweite; 45 Ludlockite, 45 Ludlockite-(Pb)*; 50 Paulmooreite, 55 Stibivanite, 60 Chadwickite

- o 04.JB Arsenites, antimonites, bismuthites; with additional anions, without H_2O: 05 Fetiasite, 10 Manganarsite, 15 Magnussonite, 20 Armangite, 25 Nanlingite, 30 Asbecasite, 35 Stenhuggarite, 40 Trigonite, 45 Finnemanite, 50 Gebhardite; 55 Derbylite, 55 Graeserite, 55 Tomichite; 60 Hemloite, 65 Freedite, 70 Georgiadesite, 75 Ekatite
- o 04.JC Arsenites, antimonites, bismuthites; without additional anions, with H_2O: 05 Cafarsite, 10 Lazarenkoite, 15 Rouseite, 20 Vajdakite
- o 04.JD Arsenites, antimonites, bismuthites; with additional anions, with H_2O: 05 Nealite, 05 Nealite-(H2O)*; 10 Seelite-1, 10 Seelite-2; 15 Tooeleite
- o 04.JE Sulfites: 05 Gravegliaite, 10 Hannebachite, 15 Orschallite, 20 Scotlandite
- o 04.JF Selenites without additional anions, without H_2O: 05 Molybdomenite
- o 04.JG Selenites with additional anions, without H_2O: Prewittite; 05 Parageorgbokiite, 05 Georgbokiite; 10 Chloromenite, 15 Sophiite, 20 Ilinskite, 25 Francisite, 30 Derriksite, 35 Burnsite, 40 Allochalcoselite
- o 04.JH Selenites without additional anions, with H_2O: 05 Chalcomenite; 10 Ahlfeldite, 10 Clinochalcomenite, 10 Cobaltomenite; 15 Mandarinoite, 20 Orlandiite, 25 Larisaite
- o 04.JJ Selenites with additional anions, with H_2O: 05 Marthozite, 10 Guilleminite, 15 Piretite, 20 Demesmaekerite, 25 Haynesite
- o 04.JK Tellurites without additional anions, without H_2O: 05 Walfordite, 05 Winstanleyite, 10 Spiroffite, 10 Zincospiroffite, 15 Balyakinite, 20 Rajite, 25 Carlfriesite, 30 Denningite, 35 Chekhovichite, 40 Smirnite, 45 Choloalite, 50 Fairbankite, 55 Plumbotellurite, 60 Magnolite, 65 Moctezumite, 70 Schmitterite, 75 Cliffordite
- o 04.JL Tellurites with additional anions, without H_2O: 05 Rodalquilarite, 10 Mackayite, 15 Mroseite, 20 Pingguite, 25 Tlapallite, 30 Girdite

- 04.JM Tellurites without additional anions, with H_2O: 05 Keystoneite, 05 Kinichilite, 05 Zemannite, 10 Emmonsite, 15 Graemite, 20 Teineite
- 04.JN Tellurites with additional anions, with H_2O: 05 Sonoraite, 10 Poughite; 15 Cesbronite, 15 Cesbronite-x; 20 Eztlite, 25 Oboyerite, 30 Juabite

- 04.K Iodates
  - 04.KA Iodates without additional anions, without H_2O: 05 Lautarite
  - 04.KB Iodates with additional anions, without H_2O: 05 Salesite, 10 Schwartzembergite, 15 Seeligerite
  - 04.KC Iodates without additional anions, with H_2O: 05 Bellingerite, 10 Bruggenite
  - 04.KD Iodates with additional anions, with H_2O: 05 Dietzeite, 10 George-ericksenite
- 04.X Unclassified Strunz Oxides (Arsenites, antimonites, bismuthites, sulfites, selenites, tellurites, iodates)
  - 04.XX Unknown: 00 Methane hydrate-I, 00 Methane hydrate-II, 00 Methane hydrate-H

Carbonate Minerals

Carbonate minerals are those minerals containing the carbonate ion: CO_3^{2-}.

Carbonate Divisions Anhydrous Carbonates

Figure 2 : Rhodochrosite, Sweet Home Mine, Alma, Colorado-5.2 x 4.2 x 2.3 cm

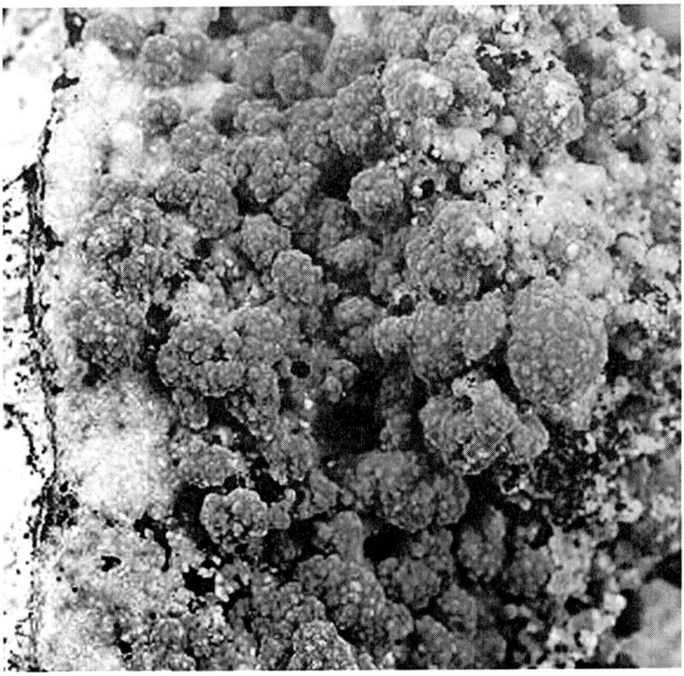

Figure 3 : Smithsonite, Silver Bill Mine, Dragoon Mts, Cochise County, Arizona-4.8 x 4.1 x 2.4 cm

- Calcite group: Trigonal
  - o Calcite $CaCO_3$
  - o Gaspeite $(Ni,Mg,Fe^{2+})CO_3$
  - o Magnesite $MgCO_3$
  - o Otavite $CdCO_3$
  - o Rhodochrosite $MnCO_3$
  - o Siderite $FeCO_3$
  - o Smithsonite $ZnCO_3$
  - o Sphaerocobaltite $CoCO_3$
- Aragonite group: Orthorhombic
  - o Aragonite $CaCO_3$
  - o Cerussite $PbCO_3$
  - o Strontianite $SrCO_3$
  - o Witherite $BaCO_3$
  - o Rutherfordine UO_2CO_3
  - o Natrite Na_2CO_3

Anhydrous carbonates with compound formulas

Figure 4 : Dolomite with calcite and chalcopyrite from the Picher Field, Tri-State district, Cherokee County, Kansas; 12.0 x 9.7 x 4.3 cm

- Dolomite group: Trigonal
  - o Ankerite $CaFe(CO_3)_2$
  - o Dolomite $CaMg(CO_3)_2$
  - o Minrecordite $CaZn(CO_3)_2$
  - o Barytocite $BaCa(CO_3)_2$

Carbonates with hydroxyl or halogen

- Carbonate with hydroxide: Monoclinic
  - o Azurite $Cu_3(CO_3)_2(OH)_2$
  - o Hydrocerussite $Pb_3(CO_3)_2(OH)_2$
  - o Malachite $Cu_2CO_3(OH)_2$
  - o Rosasite $(Cu,Zn)_2CO_3(OH)_2$
  - o Phosgenite $Pb_2(CO_3)Cl_2$
  - o Hydrozincite $Zn_5(CO_3)_2(OH)_6$
  - o Aurichalcite $(Zn,Cu)_5(CO_3)_2(OH)_6$

Hydrated Carbonates

- Ikaite $CaCO_3{\cdot}6(H_2O)$
- Lansfordite $MgCO_3{\cdot}5(H_2O)$
- Monohydrocalcite $CaCO_3{\cdot}H_2O$

- Natron $Na_2CO_3 \cdot 10(H_2O)$
- Zellerite $Ca(UO_2)(CO_3)_2 \cdot 5(H_2O)$.

The carbonate class in both the Dana and the Strunz classification systems include the nitrates.

Nickel˘Strunz Classification-05-Carbonates

IMA-CNMNC proposes a new hierarchical scheme. This list uses the Classification of Nickel–Strunz (10 ed, pending publication).

- Abbreviations:
  - "*"-discredited (IMA/CNMNC status).
  - "?"-questionable/doubtful (IMA/CNMNC status).
  - "REE"-Rare-earth element (Sc, Y, La, Ce, Pr, Nd, Pm, Sm, Eu, Gd, Tb, Dy, Ho, Er, Tm, Yb, Lu)
  - "PGE"-Platinum-group element (Ru, Rh, Pd, Os, Ir, Pt)
  - 03.C Aluminofluorides, 06 Borates, 08 Vanadates (04.H V Vanadates), 09 Silicates:
    - Neso: insular
    - Soro: grouping
    - Cyclo: ring
    - Ino: chain
    - Phyllo: sheet
    - Tekto: three-dimensional framework
- Nickel–Strunz code scheme: NN.XY.##x
  - NN: Nickel–Strunz mineral class number
  - X: Nickel–Strunz mineral division letter
  - Y: Nickel–Strunz mineral family letter
  - ##x: Nickel–Strunz mineral/group number, x add-on letter.

Class: Carbonates

- 05.A Carbonates without additional anions, without H_2O
  - 05.AA Alkali Carbonates: 05 Zabuyelite; 10 Gregoryite, 10 Natrite; 15 Nahcolite, 20 Kalicinite, 25 Teschemacherite, 30 Wegscheiderite
  - 05.AB Alkali-earth (and other M^{2+}) carbonates: 05 Calcite, 05 Gaspeite, 05 Magnesite, 05 Rhodochrosite, 05 Otavite, 05 Sphaerocobaltite, 05 Siderite, 05 Smithsonite; 10 Ankerite, 10 Dolomite, 10 Kutnohorite, 10 Minrecordite;

15 Cerussite, 15 Aragonite, 15 Strontianite, 15 Witherite; 20 Vaterite, 25 Huntite, 30 Norsethite, 35 Alstonite; 40 Olekminskite, 40 Paralstonite; 45 Barytocalcite, 50 Carbocernaite, 55 Benstonite, 60 Juangodoyite

- o 05.AC Alkali and alkali-earth carbonates: 05 Eitelite, 10 Nyerereite, 10 Natrofairchildite, 10 Zemkorite; 15 Butschliite, 20 Fairchildite, 25 Shortite; 30 Sanromanite, 30 Burbankite, 30 Calcioburbankite, 30 Khanneshite
- o 05.AD With rare-earth elements (REE): 05 Sahamalite-(Ce); 15 Remondite-(Ce), 15 Petersenite-(Ce), 15 Remondite-(La); 20 Paratooite-(La)

- 05.B Carbonates with additional anions, without H_2O
  - o 05.BA With Cu, Co, Ni, Zn, Mg, Mn: 05 Azurite, 10 Chukanovite, 10 Malachite, 10 Georgeite, 10 Pokrovskite, 10 Nullaginite, 10 Glaukosphaerite, 10 Mcguinnessite, 10 Kolwezite, 10 Rosasite, 10 Zincrosasite; 15 Aurichalcite, 15 Hydrozincite; 20 Holdawayite, 25 Defernite; 30 Loseyite, 30 Sclarite
  - o 05.BB With alkalies, etc.: 05 Barentsite, 10 Dawsonite, 15 Tunisite, 20 Sabinaite
  - o 05.BC With alkali-earth cations: 05 Brenkite, 10 Rouvilleite, 15 Podlesnoite
  - o 05.BD With rare-earth elements (REE): 05 Cordylite-(Ce), 05 Lukechangite-(Ce); 10 Kukharenkoite-(La), 10 Kukharenkoite-(Ce), 10 Zhonghuacerite-(Ce); 15 Cebaite-(Nd), 15 Cebaite-(Ce); 20a Bastnasite-(Ce), 20a Bastnasite-(La), 20a Bastnasite-(Y), 20a Hydroxylbastnasite-(Ce), 20a Hydroxylbastnasite-(La), 20a Hydroxylbastnasite-(Nd), 20a Thorbastnasite, 20b Parisite-(Nd), 20b Parisite-(Ce), 20c Synchysite-(Ce), 20c Synchysite-(Nd), 20c Synchysite-(Y), 20d Rontgenite-(Ce); 25 Horvathite-(Y), 30 Qaqarssukite-(Ce), 35 Huanghoite-(Ce)
  - o 05.BE With Pb, Bi: 05 Shannonite, 10 Hydrocerussite, 15 Plumbonacrite, 20 Phosgenite, 25 Bismutite, 30 Kettnerite, 35 Beyerite
  - o 05.BF With (Cl), SO_4, PO_4, TeO_3: 05 Northupite, 05 Ferrotychite, 05 Manganotychite, 05 Tychite; 10 Bonshtedtite, 10 Crawfordite, 10 Bradleyitev, 10 Sidorenkite, 15 Daqingshanite-(Ce), 20 Reederite-(Y), 25

Mineevite-(Y), 30 Brianyoungite, 35 Philolithite; 40 Macphersonitev, 40 Susannite, 40 Leadhillite

- 05.C Carbonates without Additional Anions, with H_2O
  - o 05.CA With medium-sized cations: 05 Nesquehonite, 10 Lansfordite, 15 Barringtonite, 20 Hellyerite
  - o 05.CB With large cations (alkali and alkali-earth carbonates): 05 Thermonatrite, 10 Natron, 15 Trona, 20 Monohydrocalcite, 25 Ikaite, 30 Pirssonite, 35 Gaylussite, 40 Chalconatronite, 45 Baylissite, 50 Tuliokite
  - o 05.CC With rare-earth elements (REE): 05 Donnayite-(Y), 05 Mckelveyite-(Nd)*, 05 Mckelveyite-(Y), 05 Weloganite; 10 Tengerite-(Y), 15 Lokkaite-(Y); 20 Shomiokite-(Y), 20 IMA2008-069; 25 Calkinsite-(Ce), 25 Lanthanite-(Ce), 25 Lanthanite-(La), 25 Lanthanite-(Nd); 30 Adamsite-(Y), 35 Decrespignyite-(Y), 40 Galgenbergite-(Ce), 45 Ewaldite, 50 Kimuraite-(Y)
- 05.D Carbonates with Additional Anions, with H_2O
  - o 05.DA With medium-sized cations: 05 Dypingite, 05 Giorgiosite, 05 Hydromagnesite, 05 Widgiemoolthalite; 10 Artinite, 10 Chlorartinite; 15 Otwayite, 20 Kambaldaite, 25 Callaghanite, 30 Claraite; 35 Hydroscarbroite, 35 Scarbroite; 40 Charmarite-3T, 40 Charmarite-2H, 40 Caresite, 40 Quintinite-2H, 40 Quintinite-3T; 45 Brugnatellite, 45 Barbertonite, 45 Chlormagaluminite, 45 Zaccagnaite, 45 Manasseite, 45 Sjogrenite; 50 Desautelsite, 50 Comblainite, 50 Hydrotalcite, 50 Pyroaurite, 50 Reevesite, 50 Stichtite, 50 Takovite; 55 Coalingite, 60 Karchevskyite, 65 Indigirite, 70 Zaratite
  - o 05.DB With large and medium-sized cations: 05 Alumohydrocalcite, 05 Para-alumohydrocalcite, 05 Nasledovite; 10 Dresserite, 10 Dundasite, 10 Strontiodresserite, 10 Petterdite, 10 Kochsandorite; 15 Hydrodresserite, 20 Schuilingite-(Nd), 25 Sergeevite, 30 Szymanskiite, 35 Montroyalite
  - o 05.DC With large cations: 05 Ancylite-(Ce), 05 Ancylite-(La), 05 Gysinite-(Nd), 05 Calcioancylite-(Ce), 05 Calcioancylite-(Nd), 05 Kozoite-(La), 05 Kozoite-(Nd); 10 Kamphaugite-(Y), 15 Sheldrickite, 20 Thomasclarkite-(Y), 25 Peterbaylissite, 30 Clearcreekite, 35 Niveolanite

- 05.E Uranyl Carbonates
  - 05.EA $UO_2:CO_3$ > 1:1: 10 Urancalcarite, 15 Wyartite, 20 Oswaldpeetersite, 25 Roubaultite, 30 Kamotoite-(Y), 35 Sharpite
  - 05.EB $UO_2:CO_3$ = 1:1: 05 Rutherfordine, 10 Blatonite, 15 Joliotite, 20 Bijvoetite-(Y)
  - 05.EC $UO_2:CO_3$ < 1:1-1:2: 05 Fontanite; 10 Metazellerite, 10 Zellerite
  - 05.ED $UO_2:CO_3$ = 1:3: 05 Bayleyite, 10 Swartzite, 15 Albrechtschraufite, 20 Liebigite, 25 Rabbittite, 30 Andersonite, 35 Grimselite, 40 Widenmannite, 45 Znucalite, 50 Cejkaite
  - 05.EE $UO_2:CO_3$ = 1:4: 05 Voglite, 10 Shabaite-(Nd)
  - 05.EF $UO_2:CO_3$ = 1:5: 05 Astrocyanite-(Ce)
  - 05.EG With SO_4 or SiO_4: 05 Schrockingerite, 10 Lepersonnite-(Gd).

Class: Nitrates

- 05.N Nitrates
  - 05.NA Without OH or H_2O: 05 Nitratine, 10 Niter, 15 Gwihabaite, 20 Nitrobarite
  - 05.NB With OH: 05 Gerhardtite, 05 Rouaite
  - 05.NC With H_2O: 05 Nitromagnesite, 10 Nitrocalcite
  - 05.ND With OH (etc.) and H_2O: 05 Likasite, 10 Mbobomkulite, 15 Hydrombobomkulite, 20 Sveite.

Borate Minerals

The borate minerals are minerals which contain a borate anion group. The borate (BO_3) units may be polymerised similar to the SiO_4 unit of the silicate mineral class. This results in B_2O_5, B_3O_6, B_2O_4 anions as well as more complex structures which include hydroxide or halogen anions. The $[B(O,OH)_4]^-$anion exists too.

Many borate minerals, such as borax, colemanite, and ulexite, are salts: soft, readily soluble, and found in evaporite contexts. However, some, such as boracite, are hard and resistant to weathering, more similar to the silicates.

There are over 100 different borate minerals. Borate minerals include:

- Kernite $Na_2B_4O_6(OH)_3{\cdot}3(H_2O)$
- Borax $Na_2B_4O_5(OH)_4{\cdot}8(H_2O)$
- Ulexite $NaCaB_5O_6(OH)_6{\cdot}5(H_2O)$
- Colemanite $CaB_3O_4(OH)_3{\cdot}(H_2O)$
- Boracite $Mg_3B_7O_{13}Cl$

Nickel˘Strunz Classification-06-Borates

IMA-CNMNC proposes a new hierarchical scheme. This list uses it to modify the Classification of Nickel–Strunz (10 ed, pending publication).

- Abbreviations:
  - o "*"-discredited (IMA/CNMNC status).
  - o "?"-questionable/doubtful (IMA/CNMNC status).
  - o "REE"-Rare-earth element (Sc, Y, La, Ce, Pr, Nd, Pm, Sm, Eu, Gd, Tb, Dy, Ho, Er, Tm, Yb, Lu)
  - o "PGE"-Platinum-group element (Ru, Rh, Pd, Os, Ir, Pt)
  - o 03.C Aluminofluorides, 06 Borates, 08 Vanadates (04.H V Vanadates), 09 Silicates:
    - * Neso: insular
    - * Soro: grouping
    - * Cyclo: ring
    - * Ino: chain
    - * Phyllo: sheet
    - * Tekto: three-dimensional framework
- Nickel–Strunz code scheme: NN.XY.##x
  - o NN: Nickel–Strunz mineral class number
  - o X: Nickel–Strunz mineral division letter
  - o Y: Nickel–Strunz mineral family letter
  - o ##x: Nickel–Strunz mineral/group number, x add-on letter.

Class: Borates

- 06. Alfredstelznerite
- 06.A Monoborates
  - o 06.AA BO_3, without additional anions; 1(D): 05 Sassolite; 15 Nordenskioldine, 15 Tusionite; 35 Jimboite, 35 Kotoite; 40 Takedaite

- 06.AB BO_3, with additional anions; 1(D) + OH, etc.: 05 Hambergite, 10 Berborite, 15 Jeremejevite; 20 Warwickite, 20 Yuanfuliite; 25 Karlite; 30 Azoproite, 30 Bonaccordite, 30 Fredrikssonite, 30 Ludwigite, 30 Vonsenite; 35 Pinakiolite; 40 Blatterite, 40 Orthopinakiolite, 40 Takeuchiite, 40 Chestermanite; 45 Hulsite, 45 Magnesiohulsite, 45 Aluminomagnesiohulsite; 50 Hydroxylborite, 50 Fluoborite; 55 Shabynite, 55 Wightmanite; 60 Gaudefroyite, 65 Sakhaite, 70 Harkerite; 75 IMA2008-060, 75 Pertsevite; 80 Jacquesdietrichite, 85 Painite
- 06.AC $B(O,OH)_4$, without and with additional anions; 1(T), 1(T)+OH, etc.: 05 Sinhalite, 10 Pseudosinhalite; 15 Behierite, 15 Schiavinatoite; 20 Frolovite, 25 Hexahydroborite, 30 Henmilite, 35 Bandylite, 40 Teepleite, 45 Moydite-(Y), 50 Carboborite, 55 Sulfoborite, 60 Luneburgite, 65 Seamanite, 70 Cahnite

- 06.H Unclassified Borates
  - 06.HA Unclassified borates: 05 Chelkarite, 10 Braitschite-(Ce), 15 Satimolite, 20 Iquiqueite, 25 Wardsmithite, 30 Korzhinskite, 35 Halurgite, 40 Ekaterinite, 45 Vitimite, 50 Canavesite, 55 Qilianshanite

Subclass: Nesoborates

- 06.BA Neso-diborates with double triangles $B_2(O,OH)_5$; 2(2D); 2(2D) + OH, etc.: 05 Suanite; 10 Clinokurchatovite, 10 Kurchatovite; 15 Sussexite, 15 Szaibelyite; 20 Wiserite
- 06.BB Neso-diborates with double tetrahedra $B_2O(OH)_6$; 2(2T): 05 Pinnoite, 10 Pentahydroborite
- 06.CA Neso-triborates: 10 Ameghinite, 15 Inderite, 20 Kurnakovite, 25 Inderborite, 30 Meyerhofferite, 35 Inyoite, 40 Solongoite, 45 Peprossiite-(Ce), 50 Nifontovite, 55 Olshanskyite
- 06.DA Neso-tetraborates: 10 Borax, 15 Tincalconite, 20 Hungchaoite; 25 Fedorovskite, 25 Roweite; 30 Hydrochlorborite, 35 Uralborite; 40 Numanoite, 40 Borcarite
- 06.EA Neso-pentaborates: 05 Sborgite; 10 Ramanite-(Rb), 10 Ramanite-(Cs), 10 Santite; 15 Ammonioborite, 25 Ulexite
- 06.FA Neso-hexaborates: 05 Aksaite, 10 Mcallisterite, 15 Admontite, 20 Rivadavite, 25 Teruggite

Subclass: Inoborates

- 06.BC Ino-diborates with triangles and/or tetrahedra: 10 Calciborite, 10 Aldzhanite*; 15 Vimsite; 20 Sibirskite, 20 Parasibirskite
- 06.CB Ino-triborates: 10 Colemanite, 15 Hydroboracite, 20 Howlite, 25 Jarandolite
- 06.DB Ino-tetraborates: 05 Kernite
- 06.EB Ino-pentaborates: 05 Larderellite, 10 Ezcurrite, 15 Probertite, 20 Tertschite, 25 Priceite
- 06.FB Ino-hexaborates: 05 Aristarainite, 10 Kaliborite.

Subclass: Phylloborates

- 06.CC Phyllo-triborates: 05 Johachidolite
- 06.EC Phyllo-pentaborates: 05 Biringuccite, 05 Nasinite; 10 Gowerite; 15 Veatchite, 15 Veatchite-A, 15 Veatchite-p; 20 Volkovskite, 25 Tuzlaite, 30 Heidornite, 35 Brianroulstonite
- 06.FC Phyllo-hexaborates: 05 Nobleite, 05 Tunellite, 05 Balavinskite?; 10 Strontioborite; 15 Ginorite, 15 Strontioginorite; 20 Fabianite
- 06.GB Phyllo-nonborates, etc.: 05 Studenitsite, 10 Penobsquisite, 15 Preobrazhenskite, 20 Walkerite

Subclass: Tektoborates

- 06.BD Tektodiborates with tetrahedra: 05 Santarosaite
- 06.DD Tekto-tetraborates: 05 Diomignite
- 06.ED Tekto-pentaborates: 05 IMA2007-047, 05 Tyretskite, 05 Hilgardite, 05 Kurgantaite
- 06.GA Tekto-heptaborates: 05 Boracite, 05 Chambersite, 05 Ericaite; 10 Congolite, 10 Trembathite
- 06.GC Tekto-dodecaborates: 05 Rhodizite, 05 Londonite
- 06.GD Mega-tektoborates: 05 Ruitenbergite, 05 Pringleite; 10 Metaborite.

Sulfate Minerals

The sulfate minerals are a class of minerals which include the sulfate ion (SO_4^{2-}) within their structure. The sulfate minerals occur commonly in primary evaporite depositional environments, as gangue minerals in hydrothermal veins and as secondary minerals in the oxidizing zone of sulfide mineral deposits. The chromate and manganate minerals have a similar structure and are often included with the sulfates in mineral classification systems.

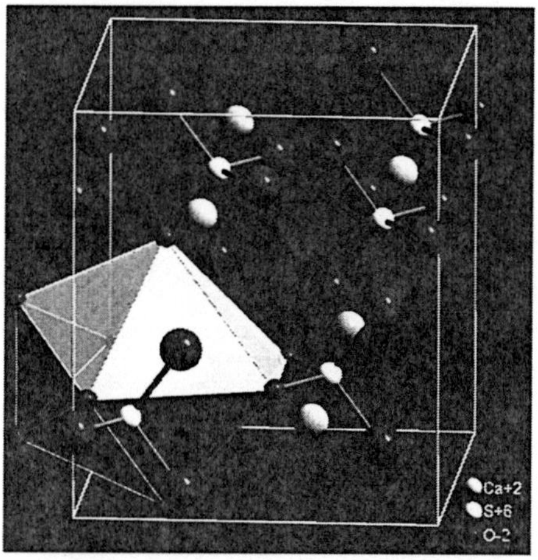

Figure 5: Anhydrite crystal structure

Figure 6 : Barite with cerussite

Sulfate minerals include:

- Anhydrous sulfates
  - o Barite $BaSO_4$
  - o Celestite $SrSO_4$
  - o Anglesite $PbSO_4$
  - o Anhydrite $CaSO_4$
  - o Hanksite $Na_{22}K(SO_4)_9(CO_3)_2Cl$
- Hydroxide and hydrous sulfates
  - o Gypsum $CaSO_4{\cdot}2H_2O$
  - o Chalcanthite $CuSO_4{\cdot}5H_2O$
  - o Kieserite $MgSO_4{\cdot}H_2O$
  - o Starkeyite $MgSO_4{\cdot}4H_2O$

- o Hexahydrite $MgSO_4 \cdot 6H_2O$
- o Epsomite $MgSO_4 \cdot 7H_2O$
- o Meridianiite $MgSO_4 \cdot 11H_2O$
- o Melanterite $FeSO_4 \cdot 7H_2O$
- o Antlerite $Cu_3SO_4(OH)_4$
- o Brochantite $Cu_4SO_4(OH)_6$
- o Alunite $KAl_3(SO_4)_2(OH)_6$
- o Jarosite $KFe_3(SO_4)_2(OH)_6$

Nickel˘Strunz Classification-07-Sulfates

IMA-CNMNC proposes a new hierarchical scheme (Mills et al., 2009). This list uses it to modify the Classification of Nickel–Strunz (10 ed, pending publication).

- Abbreviations:
  - o "*"-discredited (IMA/CNMNC status).
  - o "?"-questionable/doubtful (IMA/CNMNC status).
  - o "REE"-Rare-earth element (Sc, Y, La, Ce, Pr, Nd, Pm, Sm, Eu, Gd, Tb, Dy, Ho, Er, Tm, Yb, Lu)
  - o "PGE"-Platinum-group element (Ru, Rh, Pd, Os, Ir, Pt)
  - o 03.C Aluminofluorides, 06 Borates, 08 Vanadates (04.H V Vanadates), 09 Silicates:
    - * Neso: insular
    - * Soro: grouping
    - * Cyclo: ring
    - * Ino: chain
    - * Phyllo: sheet
    - * Tekto: three-dimensional framework
- Nickel–Strunz code scheme: NN.XY.##x
  - o NN: Nickel–Strunz mineral class number
  - o X: Nickel–Strunz mineral division letter
  - o Y: Nickel–Strunz mineral family letter
  - o ##x: Nickel–Strunz mineral/group number, x add-on letter.

Class: Sulfates, Selenates, Tellurates

- 07.A Sulfates (selenates, etc.) without Additional Anions, without H_2O

- o 07.AB With medium-sized cations: 05 Millosevichite, 05 Mikasaite; 10 Chalcocyanite, 10 Zincosite*
- o 07.AC With medium-sized and large cations: IMA2008-029, 05 Vanthoffite; 10 Efremovite, 10 Manganolangbeinite, 10 Langbeinite; 15 Eldfellite, 15 Yavapaiite; 20 Godovikovite, 20 Sabieite; 25 Thenardite, 35 Aphthitalite
- o 07.AD With only large cations: 05 Arcanite, 05 Mascagnite; 10 Mercallite, 15 Misenite, 20 Letovicite, 25 Glauberite, 30 Anhydrite; 35 Anglesite, 35 Barite, 35 Celestine, 35 Radiobarite*, 35 Olsacherite; 40 Kalistrontite, 40 Palmierite

- 07.B Sulfates (selenates, etc.) with Additional Anions, without H_2O
  - o 07.BB With medium-sized cations: 05 Caminite, 10 Hauckite, 15 Antlerite, 20 Dolerophanite, 25 Brochantite, 30 Vergasovaite, 35 Klebelsbergite, 40 Schuetteite, 45 Paraotwayite, 50 Xocomecatlite, 55 Pauflerite
  - o 07.BC With medium-sized and large cations: 05 Dansite; 10 Alunite, 10 Ammonioalunite, 10 Ammoniojarosite, 10 Beaverite, 10 Argentojarosite, 10 Huangite, 10 Dorallcharite, 10 Jarosite, 10 Hydroniumjarosite, 10 Minamiite, 10 Natrojarosite, 10 Natroalunite, 10 Osarizawaite, 10 Plumbojarosite, 10 Walthierite, 10 Schlossmacherite; 15 Yeelimite; 20 Atlasovite, 20 Nabokoite; 25 Chlorothionite; 30 Fedotovite, 30 Euchlorine; 35 Kamchatkite, 40 Piypite; 45 Klyuchevskite-Duplicate, 45 Klyuchevskite, 45 Alumoklyuchevskite; 50 Caledonite, 55 Wherryite, 60 Mammothite; 65 Munakataite, 65 Schmiederite, 65 Linarite; 70 Chenite, 75 Krivovichevite
  - o 07.BD With only large cations: 05 Sulphohalite; 10 Galeite, 10 Schairerite; 15 Kogarkoite; 20 Cesanite, 20 Caracolite; 25 Burkeite, 30 Hanksite, 35 Cannonite, 40 Lanarkite, 45 Grandreefite, 50 Itoite, 55 Chiluite, 60 Hectorfloresite, 65 Pseudograndreefite, 70 Sundiusite
- 07.C Sulfates (selenates, etc.) without Additional Anions, with H_2O
  - o 07.CB With only medium-sized cations: 05 Gunningite, 05 Dwornikite, 05 Kieserite, 05 Szomolnokite, 05 Szmikite, 05 Poitevinite, 05 Cobaltkieserite; 07 Sanderite, 10 Bonattite, 15 Boyleite, 15 Aplowite, 15 Ilesite, 15 Rozenite, 15

Starkeyite, 15 IMA2002-034; 20 Chalcanthite, 20 Jokokuite, 20 Pentahydrite, 20 Siderotil; 25 Bianchite, 25 Ferrohexahydrite, 25 Chvaleticeite, 25 Hexahydrite, 25 Moorhouseite, 25 Nickelhexahydrite; 30 Retgersite; 35 Bieberite, 35 Boothite, 35 Mallardite, 35 Melanterite, 35 Zincmelanterite, 35 Alpersite; 40 Epsomite, 40 Goslarite, 40 Morenosite; 45 Alunogen, 45 Meta-alunogen; 50 Coquimbite, 50 Paracoquimbite; 55 Rhomboclase, 60 Kornelite, 65 Quenstedtite, 70 Lausenite; 75 Lishizhenite, 75 Romerite; 80 Ransomite; 85 Bilinite, 85 Apjohnite, 85 Dietrichite, 85 Halotrichite, 85 Pickeringite, 85 Redingtonite, 85 Wupatkiite; 90 Meridianiite, 95 Caichengyunite

- 07.CC With medium-sized and large cations: 05 Krausite, 10 Tamarugite; 15 Mendozite, 15 Kalinite; 20 Lonecreekite, 20 Alum-(K), 20 Alum-(Na), 20 Lanmuchangite, 20 Tschermigite; 25 Pertlikite, 25 Monsmedite?, 25 Voltaite, 25 Zincovoltaite; 30 Krohnkite, 35 Ferrinatrite, 40 Goldichite, 45 Loweite; 50 Blodite, 50 Changoite, 50 Nickelblodite; 55 Mereiterite, 55 Leonite; 60 Boussingaultite, 60 Cyanochroite, 60 Mohrite, 60 Picromerite, 60 Nickelboussingaultite; 65 Polyhalite; 70 Leightonite, 75 Amarillite, 80 Konyaite, 85 Wattevilleite
- 07.CD With only large cations: 05 Matteuccite, 10 Mirabilite, 15 Lecontite, 20 Hydroglauberite, 25 Eugsterite, 30 Gorgeyite; 35 Koktaite, 35 Syngenite; 40 Gypsum, 45 Bassanite, 50 Zircosulfate, 55 Schieffelinite, 60 Montanite, 65 Omongwaite

- 07.D Sulfates (selenates, etc.) with additional anions, with H_2O
  - 07.DB With only medium-sized cations; insular octahedra and finite groups: 05 Svyazhinite, 05 Aubertite, 05 Magnesioaubertite; 10 Rostite, 10 Khademite; 15 Jurbanite; 20 Minasragrite, 20 Anorthominasragrite, 20 Orthominasragrite; 25 Bobjonesite; 30 Amarantite, 30 Hohmannite, 30 Metahohmannite; 35 Aluminocopiapite, 35 Copiapite, 35 Calciocopiapite, 35 Cuprocopiapite, 35 Ferricopiapite, 35 Magnesiocopiapite, 35 Zincocopiapite
  - 07.DC With only medium-sized cations; chains of corner-sharing octahedra: 05 Aluminite, 05 Meta-aluminite; 10

Butlerite, 10 Parabutlerite; 15 Fibroferrite, 20 Xitieshanite; 25 Botryogen, 25 Zincobotryogen; 30 Chaidamuite, 30 Guildite

- o 07.DD With only medium-sized cations; sheets of edge-sharing octahedra: 05 Basaluminite?, 05 Felsobanyaite, 07.5 Kyrgyzstanite, 08.0 Zn-Schulenbergite; 10 Langite, 10 Posnjakite, 10 Wroewolfeite; 15 Spangolite, 20 Ktenasite, 25 Christelite; 30 Campigliaite, 30 Devilline, 30 Orthoserpierite, 30 Niedermayrite, 30 Serpierite; 35 Motukoreaite, 35 Mountkeithite, 35 Glaucocerinite, 35 Honessite, 35 Hydrowoodwardite, 35 Hydrohonessite, 35 Shigaite, 35 Natroglaucocerinite, 35 Wermlandite, 35 Nikischerite, 35 Zincaluminite, 35 Woodwardite, 35 Carrboydite, 35 Zincowoodwardite, 35 Zincowoodwardite-3R, 35 Zincowoodwardite-1T; 40 Lawsonbauerite, 40 Torreyite, 45 Mooreite, 50 Namuwite, 55 Bechererite, 60 Ramsbeckite, 65 Vonbezingite, 70 Redgillite; 75 Chalcoalumite, 75 Nickelalumite*; 80 Guarinoite, 80 Theresemagnanite, 80 Schulenbergite; 85 Montetrisaite
- o 07.DE With only medium-sized cations; unclassified: 05 Mangazeite; 10 Carbonatecyanotrichite, 10 Cyanotrichite; 15 Schwertmannite, 20 Tlalocite, 25 Utahite, 35 Coquandite, 40 Osakaite, 45 Wilcoxite, 50 Stanleyite, 55 Mcalpineite, 60 Hydrobasaluminite, 65 Zaherite, 70 Lautenthalite, 75 Camerolaite, 80 Brumadoite
- o 07.DF With large and medium-sized cations: 05 Uklonskovite, 10 Kainite, 15 Natrochalcite; 20 Metasideronatrite, 20 Sideronatrite; 25 Despujolsite, 25 Fleischerite, 25 Schaurteite, 25 Mallestigite; 30 Slavikite, 35 Metavoltine; 40 Lannonite, 40 Vlodavetsite; 45 Peretaite, 50 Gordaite, 55 Clairite, 60 Arzrunite, 65 Elyite, 70 Yecoraite, 75 Riomarinaite, 80 Dukeite, 85 Xocolatlite
- o 07.DG With large and medium sized cations; with NO_3, CO_3, $B(OH)_4$, SiO_4 or IO_3: 05 Darapskite; 10 Clinoungemachite, 10 Ungemachite, 10 Humberstonite; 15 Bentorite, 15 Charlesite, 15 Ettringite, 15 Jouravskite, 15 Sturmanite, 15 Thaumasite, 15 Carraraite, 15 Buryatite; 20 Rapidcreekite, 25 Tatarskite, 30 Nakauriite, 35

Chessexite; 40 Carlosruizite, 40 Fuenzalidaite; 45 Chelyabinskite*

- 07.E Uranyl Sulfates
  - o 07.EA Without cations: 05 Uranopilite, 05 Metauranopilite, 10 Jachymovite
  - o 07.EB With medium-sized cations: 05 Johannite, 10 Deliensite
  - o 07.EC With medium-sized and large cations: 05 Cobaltzippeite, 05 Magnesiozippeite, 05 Nickelzippeite, 05 Natrozippeite, 05 Zinc-zippeite, 05 Zippeite; 10 Rabejacite, 15 Marecottite, 20 Pseudojohannite
- 07.J Thiosulfates
  - o 07.JA Thiosulfates with Pb: 05 Sidpietersite
- 07.X Unclassified Strunz Sulfates (Selenates, Tellurates)
  - o 07.XX Unknown: 00 Aiolosite, 00 Steverustite, 00 Grandviewite, 00 IMA2009-008, 00 Adranosite, 00 Blakeite.

Class: Chromates

- 07.F Chromates
  - o 07.FA Without additional anions: 05 Tarapacaite, 10 Chromatite, 15 Hashemite, 20 Crocoite
  - o 07.FB With additional O, V, S, Cl: 05 Phoenicochroite, 10 Santanaite, 15 Wattersite, 20 Deanesmithite, 25 Edoylerite
  - o 07.FC With PO_4, AsO_4, SiO_4: 05 Vauquelinite; 10 Fornacite, 10 Molybdofornacite; 15 Hemihedrite, 15 Iranite; 20 Embreyite, 20 Cassedanneite;
  - o 07.FD Dichromates: 05 Lopezite.

Class: molybdates

- 07.G Molybdates
  - o 07.GA Without additional anions or H_2O: 05 Powellite, 05 Wulfenite
  - o 07.GB With additional anions and/or H_2O: 05 Lindgrenite, 10 Szenicsite, 30 Ferrimolybdite, 40 Obradovicite; 45 Mendozavilite, 45 Paramendozavilite
- 07.H Uranium and Uranyl Molybdates
  - o 07.HA With U^{4+}: 05 Sedovite, 10 Cousinite, 15 Moluranite
  - o 07.HB With U^{6+}: 15 Calcurmolite, 20 Tengchongite

Class: Tungstates

- 07.G Wolframates and Niobates
    - o 07.GA Without additional anions or H_2O: 05 Fergusonite-(Ce)*, 05 Fergusonite-(Nd)*, 05 Fergusonite-(Y), 05 Stolzite, 05 Scheelite; 10 Formanite-(Y), 10 Iwashiroite-(Y); 15 Paraniite-(Y)
    - o 07.GB With additional anions and/or H_2O: 15 Cuprotungstite, 20 Phyllotungstite, 25 Rankachite, 35 Anthoinite, 35 Mpororoite
- 07.H Uranium and Uranyl Wolframates
    - o 07.HA With U^{4+}: (none)
    - o 07.HB With U^{6+}: 25 Uranotungstite

Phosphate Minerals

Phosphate minerals are those minerals that contain the tetrahedrally coordinated phosphate (PO_4^{3-}) anion along with the freely substituting arsenate (AsO_4^{3-}) and vanadate (VO_4^{3-}). Chlorine (Cl^-), fluorine (F^-), and hydroxide (OH^-) anions also fit into the crystal structure.

The phosphate class of minerals is a large and diverse group, however, only a few species are relatively common.

Examples

Phosphate minerals include:

- triphylite $Li(Fe,Mn)PO_4$
- monazite $(Ce,La,Y,Th)PO_4$
- hinsdalite $PbAl_3(PO_4)(SO_4)(OH)_6$
- pyromorphite $Pb_5(PO_4)_3Cl$
- vanadinite $Pb_5(VO_4)_3Cl$
- erythrite $Co_3(AsO_4)_2 \cdot 8H_2O$
- amblygonite $LiAlPO_4F$
- lazulite $(Mg,Fe)Al_2(PO_4)_2(OH)_2$
- wavellite $Al_3(PO_4)_2(OH)_3 \cdot 5H_2O$
- turquoise $CuAl_6(PO_4)_4(OH)_8 \cdot 5H_2O$
- autunite $Ca(UO_2)_2(PO_4)_2 \cdot 10\text{-}12H_2O$
- carnotite $K_2(UO_2)_2(VO_4)_2 \cdot 3H_2O$
- phosphophyllite $Zn_2(Fe,Mn)(PO_4)_2 \bullet 4H_2O$
- struvite $(NH_4)MgPO_4 \cdot 6H_2O$

- Apatite group $Ca_5(PO_4)_3(F,Cl,OH)$
  - o hydroxylapatite $Ca_5(PO_4)_3OH$
  - o fluorapatite $Ca_5(PO_4)_3F$
  - o chlorapatite $Ca_5(PO_4)_3Cl$
  - o bromapatite
- Mitridatite group:
  - o Arseniosiderite-mitridatite series $(Ca_2(Fe^{3+})_3[(O)_2|(AsO_4)_3]\cdot 3H_2O$ — $Ca_2(Fe^{3+})_3[(O)_2|(PO_4)_3]\cdot 3H_2O)$
  - o Arseniosiderite-robertsite series $(Ca_2(Fe^{3+})_3[(O)_2|(AsO_4)_3]\cdot 3H_2O$ — $Ca_3(Mn^{3+})_4[(OH)_3|(PO_4)_2]_2\cdot 3H_2O)$

Applications

Phosphate rock is a general term that refers to rock with high concentration of phosphate minerals, most commonly of the apatite group. It is the major resource mined to produce phosphate fertilisers for the agriculture sector. Phosphate is also used in animal feed supplements, food preservatives, anti-corrosion agents, cosmetics, fungicides, ceramics, water treatment and metallurgy.

The largest use of minerals mined for their phosphate content is the production of fertilizer.

Phosphate minerals are often used for control of rust and prevention of corrosion on ferrous materials applied with electrochemical conversion coatings.

Nickel˘Strunz Classification-08-Phosphates

IMA-CNMNC proposes a new hierarchical scheme. This list uses it to modify the Classification of Nickel–Strunz (10 ed, pending publication).

- Abbreviations:
  - o "*"-discredited (IMA/CNMNC status).
  - o "?"-questionable/doubtful (IMA/CNMNC status).
  - o "REE"-Rare-earth element (Sc, Y, La, Ce, Pr, Nd, Pm, Sm, Eu, Gd, Tb, Dy, Ho, Er, Tm, Yb, Lu)
  - o "PGE"-Platinum-group element (Ru, Rh, Pd, Os, Ir, Pt)
  - o 03.C Aluminofluorides, 06 Borates, 08 Vanadates (04.H V Vanadates), 09 Silicates:
    - * Neso: insular

    * Soro: grouping
    * Cyclo: ring
    * Ino: chain
    * Phyllo: sheet
    * Tekto: three-dimensional framework

- Nickel–Strunz code scheme: NN.XY.##x
  - o NN: Nickel–Strunz mineral class number
  - o X: Nickel–Strunz mineral division letter
  - o Y: Nickel–Strunz mineral family letter
  - o ##x: Nickel–Strunz mineral/group number, x add-on letter.

Class: Phosphates

- 08.A Phosphates, etc. without additional anions, without H_2O
  - o 08.AA With small cations (some also with larger ones): 05 Berlinite, 05 Rodolicoite; 10 Beryllonite, 15 Hurlbutite, 20 Lithiophosphate, 25 Nalipoite, 30 Olympite
  - o 08.AB With medium-sized cations: 05 Farringtonite; 10 Ferrisicklerite, 10 Heterosite, 10 Natrophilite, 10 Lithiophilite, 10 Purpurite, 10 Sicklerite, 10 Simferite, 10 Triphylite; 15 Chopinite, 15 Sarcopside; 20 Beusite, 20 Graftonite
  - o 08.AC With medium-sized and large cations: 10 IMA2008-054, 10 Alluaudite, 10 Hagendorfite, 10 Ferroalluaudite, 10 Maghagendorfite, 10 Varulite, 10 Ferrohagendorfite*; 15 Bobfergusonite, 15 Ferrowyllieite, 15 Qingheiite, 15 Rosemaryite, 15 Wyllieite, 15 Ferrorosemaryite; 20 Maricite, 30 Brianite, 35 Vitusite-(Ce); 40 Olgite?, 40 Bario-olgite; 45 Ferromerrillite, 45 Bobdownsite, 45 Merrillite-(Ca)*, 45 Merrillite, 45 Merrillite-(Y)*, 45 Whitlockite, 45 Tuite, 45 Strontiowhitlockite; 50 Stornesite-(Y), 50 Xenophyllite, 50 Fillowite, 50 Chladniite, 50 Johnsomervilleite, 50 Galileiite; 55 Harrisonite, 60 Kosnarite, 65 Panethite, 70 Stanfieldite, 90 IMA2008-064
  - o 08.AD With only large cations: 05 Nahpoite, 10 Monetite, 15 Archerite, 15 Biphosphammite; 20 Phosphammite, 25 Buchwaldite; 35 Pretulite, 35 Xenotime-(Y), 35 Xenotime-(Yb); 45 Ximengite, 50 Monazite-(Ce), 50 Monazite-(La), 50 Monazite-(Nd), 50 Monazite-(Sm), 50 Brabantite?

- 08.B Phosphates, etc. with Additional Anions, without H_2O
  - 08.BA With small and medium-sized cations: 05 Vayrynenite; 10 Hydroxylherderite, 10 Herderite; 15 Babefphite
  - 08.BB With only medium-sized cations, (OH, etc.):RO_4 £ 1:1: 05 Amblygonite, 05 Natromontebrasite?, 05 Montebrasite?, 05 Tavorite; 10 Zwieselite, 10 Triplite, 10 Magniotriplite?, 10 Hydroxylwagnerite; 15 Joosteite, 15 Stanekite, 15 Triploidite, 15 Wolfeite, 15 Wagnerite; 20 Satterlyite, 20 Holtedahlite; 25 Althausite; 30 Libethenite, 30 Zincolibethenite; 35 Tarbuttite; 40 Barbosalite, 40 Hentschelite, 40 Scorzalite, 40 Lazulite; 45 Trolleite, 55 Phosphoellenbergerite; 90 Zinclipscombite, 90 Lipscombite, 90 Richellite
  - 08.BC With only medium-sized cations, (OH, etc.):RO_4 > 1:1 and < 2:1: 10 Plimerite, 10 Frondelite, 10 Rockbridgeite
  - 08.BD With only medium-sized cations, (OH, etc.):RO_4 = 2:1: 05 Pseudomalachite, 05 Reichenbachite, 10 Gatehouseite, 25 Ludjibaite
  - 08.BE With only medium-sized cations, (OH, etc.):RO_4 > 2:1: 05 Augelite, 10 Grattarolaite, 15 Cornetite, 30 Raadeite, 85 Waterhouseite
  - 08.BF With medium-sized and large cations, (OH, etc.):RO_4 < 0.5:1: 05 Arrojadite, 05 Arrojadite-(BaFe), 05 Arrojadite-(KFe), 05 Arrojadite-(NaFe), 05 Arrojadite-(SrFe), 05 Arrojadite-(KNa), 05 Arrojadite-(PbFe), 05 Arrojadite-(BaNa), 05 Fluorarrojadite-(BaNa), 05 Fluorarrojadite-(KNa), 05 Fluorarrojadite-(BaFe), 05 Ferri-arrojadite-(BaNa), 05 Dickinsonite, 05 Dickinsonite-(KNa), 05 Dickinsonite-(KMnNa), 05 Dickinsonite-(KNaNa), 05 Dickinsonite-(NaNa); 10 Samuelsonite, 15 Griphite, 20 Nabiasite
  - 08.BG With medium-sized and large cations, (OH, etc.):RO_4 = 0.5:1: 05 Bearthite, 05 Goedkenite, 05 Tsumebite; 10 Melonjosephite, 15 Tancoite
  - 08.BH With medium-sized and large cations, (OH,etc.):RO_4 = 1:1: 05 Thadeuite; 10 Lacroixite, 10 Isokite, 10 Panasqueiraite; 15 Drugmanite; 20 Bjarebyite, 20 Kulanite, 20 Penikisite, 20 Perloffite, 20 Johntomaite; 25 Bertossaite, 25 Palermoite; 55 Jagowerite, 60 Attakolite

- 08.BK With medium-sized and large cations, (OH, etc.): 05 Brazilianite, 15 Curetonite, 25 Lulzacite
- 08.BL With medium-sized and large cations, (OH, etc.):RO_4 = 3:1: 05 Corkite, 05 Hinsdalite, 05 Orpheite, 05 Woodhouseite, 05 Svanbergite; 10 Kintoreite, 10 Benauite, 10 Crandallite, 10 Goyazite, 10 Springcreekite, 10 Gorceixite; 10 Lusungite?, 10 Plumbogummite, 10 Ferrazite?; 13 Eylettersite, 13 Florencite-(Ce), 13 Florencite-(La), 13 Florencite-(Nd), 13 Waylandite, 13 Zairite; 15 Viitaniemiite, 20 Kuksite, 25 Pattersonite
- 08.BM With medium-sized and large cations, (OH, etc.):RO_4 = 4:1: 10 Paulkellerite, 15 Brendelite
- 08.BN With only large cations, (OH, etc.):RO_4 = 0.33:1: 05 IMA2008-068, 05 Phosphohedyphane, 05 IMA2008-009, 05 Alforsite, 05 Apatite*, 05 Apatite-(CaOH), 05 Apatite-(CaCl), 05 Apatite-(CaF), 05 Apatite-(SrOH), 05 Apatite-(CaOH)-M, Carbonate-fluorapatite?, 05 Carbonate-hydroxylapatite?, 05 Belovite-(Ce), 05 Belovite-(La), 05 Fluorcaphite, 05 Pyromorphite, 05 Hydroxylpyromorphite, 05 Deloneite-(Ce), 05 Kuannersuite-(Ce), 10 Arctite
- 08.BO With only large cations, (OH, etc.):RO_4 1:1: 05 Nacaphite, 10 Petitjeanite, 15 Smrkovecite, 25 Heneuite, 30 Nefedovite, 40 Artsmithite

• 08.C Phosphates without Additional Anions, with H_2O

- 08.CA With small and large/medium cations: 05 Fransoletite, 05 Parafransoletite; 10 Ehrleite, 15 Faheyite; 20 Gainesite, 20 Mccrillisite, 20 Selwynite; 25 Pahasapaite, 30 Hopeite, 40 Phosphophyllite; 45 Parascholzite, 45 Scholzite; 65 Gengenbachite, 70 Parahopeite
- 08.CB With only medium-sized cations, RO_4:H_2O = 1:1: 05 Serrabrancaite, 10 Hureaulite
- 08.CC With only medium-sized cations, RO_4:H_2O = 1:1.5: 05 Garyansellite, 05 Kryzhanovskite, 05 Landesite, 05 Phosphoferrite, 05 Reddingite
- 08.CD With only medium-sized cations, RO_4:H_2O = 1:2: 05 Kolbeckite, 05 Metavariscite, 05 Phosphosiderite; 10 Strengite, 10 Variscite; 20 Ludlamite
- 08.CE With only medium-sized cations, RO_4:H_2O £ 1:2.5: 10 Newberyite, 20 Phosphorrosslerite; 25 Metaswitzerite, 25 Switzerite; 35 Bobierrite; 40 Arupite, 40 Baricite, 40

Vivianite, 40 Pakhomovskyite; 50 Cattiite, 55 Koninckite; 75 IMA2008-046, 75 Malhmoodite; 80 Santabarbaraite, 85 Metavivianite

- 08.CF With large and medium-sized cations, $RO_4:H_2O$ > 1:1: 05 Tassieite, 05 Wicksite, 05 Bederite; 10 Haigerachite
- 08.CG With large and medium-sized cations, $RO_4:H_2O$ = 1:1: 05 Collinsite, 05 Cassidyite, 05 Fairfieldite, 05 Messelite, 05 Hillite, (05 Uranophane-beta but Uranophane 09.AK.15); 20 Phosphogartrellite
- 08.CH With large and medium-sized cations, $RO_4:H_2O$ < 1:1: 10 Anapaite, 20 Dittmarite, 20 Niahite, 25 Francoanellite, 25 Taranakite, 30 Schertelite, 35 Hannayite, 40 Hazenite, 40 Struvite, 40 Struvite-(K), 45 Rimkorolgite, 50 Bakhchisaraitsevite, 55 IMA2008-048
- 08.CJ With only large cations: 05 Stercorite, 10 Mundrabillaite, 10 Swaknoite, 15 Nastrophite, 15 Nabaphite, 45 Brockite, 45 Grayite, 45 Rhabdophane-(Ce), 45 Rhabdophane-(La), 45 Rhabdophane-(Nd), 45 Tristramite, 50 Brushite, 50 Churchite-(Dy)*, 50 Churchite-(Nd), 50 Churchite-(Y), 50 Ardealite, 60 Dorfmanite, 70 Catalanoite, 80 Ningyoite

- 08.D Phosphates
  - 08.DA With small (and occasionally larger) cations: 05 Moraesite, 10 Footemineite, 10 Ruifrancoite, 10 Guimaraesite, 10 Roscherite, 10 Zanazziite, 10 Atencioite, 10 Greifensteinite; 15 Uralolite, 20 Weinebeneite, 25 Tiptopite, 30 Veszelyite, 35 Kipushite, 40 Spencerite, 45 Glucine
  - 08.DB With only medium-sized cations, (OH, etc.):RO_4 < 1:1: 05 Diadochite, 10 Vashegyite, 15 Schoonerite, 20 Sinkankasite, 25 Mitryaevaite, 30 Sanjuanite, 50 Giniite, 55 Sasaite, 60 Mcauslanite, 65 Goldquarryite, 70 Birchite
  - 08.DC With only medium-sized cations, (OH, etc.):RO_4 = 1:1 and < 2:1: 05 Nissonite; 15 Kunatite, 15 Earlshannonite, 15 Whitmoreite; 17 Kleemanite, 20 Bermanite, 207? Oxiberaunite*, 22 Kovdorskite; 25 Ferrostrunzite, 25 Ferristrunzite, 25 Metavauxite, 25 Strunzite; 27 Beraunite; 30 Gordonite, 30 Laueite, 30 Sigloite, 30 Paravauxite, 30 Ushkovite, 30 Ferrolaueite, 30 Mangangordonite, 30 Pseudolaueite, 30 Stewartite, 30 Kastningite, 35 Vauxite,

37 Vantasselite, 40 Cacoxenite; 45 Gormanite, 45 Souzalite; 47 Kingite; 50 Wavellite, 50 Allanpringite, 52 Kribergite, 60 Nevadaite

- o 08.DD With only medium-sized cations, (OH, etc.):RO_4 = 2:1: 15 Aheylite, 15 Chalcosiderite, 15 Faustite, 15 Planerite, 15 Turquoise; 20 Ernstite, 20 Childrenite, 20 Eosphorite
- o 08.DE With only medium-sized cations, (OH, etc.):RO_4 = 3:1: 05 Senegalite, 10 Fluellite, 20 Zapatalite, (35 Alumoakermanite, Mindat.org: 09.BB.10), 35 Aldermanite
- o 08.DF With only medium-sized cations, (OH,etc.):RO_4 > 3:1: 05 Hotsonite-VII, 05 Hotsonite-VI; 10 Bolivarite, 10 Evansite, 10 Rosieresite, 25 Sieleckiite, 40 Gladiusite
- o 08.DG With large and medium-sized cations, (OH, etc.):RO_4 < 0.5:1: 05 Sampleite
- o 08.DH With large and medium-sized cations, (OH, etc.):RO_4 < 1:1: 05 Minyulite; 10 Leucophosphite, 10 Spheniscidite, 10 Tinsleyite; 15 Kaluginite*, 15 Keckite, 15 Jahnsite-(CaMnFe), 15 Jahnsite-(CaMnMg), 15 Jahnsite-(CaMnMn), 15 Jahnsite-(MnMnMn)*, 15 Jahnsite-(CaFeFe), 15 Jahnsite-(NaFeMg), 15 Jahnsite-(CaMgMg), 15 Jahnsite-(NaMnMg), 15 Rittmannite, 15 Whiteite-(MnFeMg), 15 Whiteite-(CaFeMg), 15 Whiteite-(CaMnMg); 20 Manganosegelerite, 20 Overite, 20 Segelerite, 20 Wilhelmvierlingite, 20 Juonniite; 25 Calcioferrite, 25 Kingsmountite, 25 Montgomeryite, 25 Zodacite; 30 Lunokite, 30 Pararobertsite, 30 Robertsite, 30 Mitridatite; 35 Matveevite?, 35 Mantienneite, 35 Paulkerrite, 35 Benyacarite, 40 Xanthoxenite, 55 Englishite
- o 08.DJ With large and medium-sized cations, (OH, etc.):RO_4 = 1:l: 05 Johnwalkite, 05 Olmsteadite, 10 Gatumbaite, 20 Meurigite-Na, 20 Meurigite-K, 20 Phosphofibrite, 25 Jungite, 30 Wycheproofite, 35 Ercitite, 40 Mrazekite
- o 08.DK With large and medium-sized cations, (OH, etc.):RO_4 > 1:1 and < 2:1: 15 Matioliite, 15 IMA2008-056, 15 Dufrenite, 15 Burangaite, 15 Natrodufrenite; 20 Kidwellite, 25 Bleasdaleite, 30 Matulaite, 35 Krasnovite
- o 08.DL With large and medium-sized cations, (OH, etc.):RO_4 = 2:1: 05 Foggite; 10 Cyrilovite, 10 Millisite, 10 Wardite; 15 Petersite-(Y), 15 Calciopetersite; 25 Angastonite

- o 08.DM With large and medium-sized cations, (OH, etc.):RO_4 > 2:1: 05 Morinite, 15 Melkovite, 25 Gutsevichite?, 35 Delvauxite
- o 08.DN With only large cations: 05 Natrophosphate, 10 Isoclasite, 15 Lermontovite, 20 Vyacheslavite
- o 08.DO With CO_3, SO_4, SiO_4: 05 Girvasite, 10 Voggite, 15 Peisleyite, 20 Perhamite, 25 Saryarkite-(Y), 30 Micheelsenite, 40 Parwanite, 45 Skorpionite

- 08.E Uranyl Phosphates
  - o 08.EA UO_2:RO_4 = 1:2: 05 Phosphowalpurgite, 10 Parsonsite, 15 Ulrichite, 20 Lakebogaite
  - o 08.EB UO_2:RO_4 = 1:1: 05 Autunite, 05 Uranocircite, 05 Torbernite, 05 Xiangjiangite, 05 Saleeite; 10 Bassetite, 10 Meta-autunite, 10 Metauranocircite, 10 Metatorbernite, 10 Lehnerite, 10 Przhevalskite; 15 Chernikovite, 15 Meta-ankoleite, 15 Uramphite; 20 Threadgoldite, 25 Uranospathite, 30 Vochtenite, 35 Coconinoite, 40 Ranunculite, 45 Triangulite, 50 Furongite, 55 Sabugalite
  - o 08.EC UO_2:RO_4 = 3:2: 05 Francoisite-(Ce), 05 Francoisite-(Nd), 05 Phuralumite, 05 Upalite; 10 Kivuite?, 10 Yingjiangite, 10 Renardite, 10 Dewindtite, 10 Phosphuranylite; 15 Dumontite; 20 Metavanmeersscheite, 20 Vanmeersscheite; 25 Althupite, 30 Mundite, 35 Phurcalite, 40 Bergenite
  - o 08.ED Unclassified: 05 Moreauite, 10 Sreinite, 15 Kamitugaite
- 08.F Polyphosphates
  - o 08.FA Polyphosphates, without OH and H_2O; dimers of corner-sharing RO_4 tetrahedra: 20 Pyrocoproite*, 20 Pyrophosphite*
  - o 08.FC Polyphosphates, with H_2O only: 10 Canaphite, 20 Arnhemite*, 25 Wooldridgeite, 30 Kanonerovite
- 08.X Unclassified Strunz Phosphates
  - o 08.XX Unknown: 00 Sodium-autunite, 00 Pseudo-autunite*, 00 Cheralite-(Ce)?, 00 Laubmannite?, 00 Spodiosite?, 00 Sodium meta-autunite, 00 Kerstenite?, 00 Lewisite, 00 Coeruleolactite, 00 Viseite, 00 IMA2009-005

Chapter 3

List of Minerals

Abenakiite-(Ce)

Abenakiite-(Ce) (or IMA1991-054) is a mineral of sodium, cerium, neodymium, lanthanum, praseodymium, thorium, samarium, oxygen, sulfur, carbon, phosphorus, and silicon with a chemical formula $Na_{26}(Ce,Nd,La,Pr,Th,Sm)_6$-$[O \mid SO_2 \mid (CO_3)_6 \mid (PO_4)_6 \mid Si_6O_{18}]$. It is named after the Abenaki, an Algonquian Indian tribe of New England. Its Mohs scale rating is 4 to 5.

Abernathyite

Abernathyite is a rare hydrous uranium oxide mineral with chemical formula $K_2(UO_2)_2(AsO_4)_2{\cdot}6H_2O$. It occurs as a secondary fracture filling mineral. It crystallizes in the tetragonal system and forms yellow vitreous tabular crystals with perfect cleavage. It has a specific gravity of 3.3 and a hardness of 2.5. It was first reported from a Colorado Plateau-type uranium deposit near Moab, Utah. It is named after mine operator Jess Abernathy.

Abhurite

Abhurite is a mineral of tin, oxygen, hydrogen, and chlorine with the formula $Sn_3O(OH)_2Cl_2$. It is named after Sharm Abhur, a cove of the Red Sea. Its Mohs scale rating is 2.

Abswurmbachite

Abswurmbachite is a copper manganese silicate mineral $((Cu,Mn^{2+})Mn^{3+}{}_6O_8SiO_4)$. It was first described in 1991 and named after Irmgard Abs-Wurmbach, a German mineralogist. It crystallizes in the tetragonal system. Its Mohs scale rating is 6.5 and a specific gravity of 4.96. It has a metallic luster and its colour is jet black, with light brown streaks.

Acanthite

Acanthite, Ag_2S, crystallizes in the monoclinic system and is the stable form of silver sulfide below 173 °C. Argentite is the stable form above that temperature. As argentite cools below that temperature its cubic form is distorted to the monoclinic form of acanthite. Below 173 °C acanthite forms directly. Acanthite is the only stable form in normal air temperature.

Occurrence

Figure 1: Acanthite-Locality: Chispas Mine, Arizpe, Mun. de Arizpe, Sonora, Mexico-Scale is one inch with a rule at one cm

Acanthite is a common silver mineral in moderately low-temperature hydrothermal veins and in zones of supergene enrichment. It occurs in association with native silver, pyrargyrite, proustite, polybasite, stephanite, aguilarite, galena, chalcopyrite, sphalerite, calcite and quartz.

Acanthite was first described in 1855 for an occurrence in the Jáchymov (St Joachimsthal) District, Krušné Hory Mts (Erzgebirge), Karlovy Vary Region, Bohemia, Czech Republic. The name is from the Greek "akantha" meaning thorn or arrow, in reference to its crystal shape.

Actinolite

Actinolite is an amphibole silicate mineral with the chemical formula $Ca_2(Mg,Fe)_5Si_8O_{22}(OH)_2$.

Etymology

The name *actinolite* is derived from the Greek word *aktis*, meaning "beam" or "ray", because of the mineral's fibrous nature. (This word is also the origin of the name of the chemical element actinium.)

Mineralogy

Actinolite is an intermediate member in a solid-solution series between magnesium-rich tremolite, $Ca_2Mg_5Si_8O_{22}(OH)_2$, and iron-rich ferro-actinolite, $Ca_2Fe_5Si_8O_{22}(OH)_2$. Mg and Fe ions can be freely exchanged in the crystal structure. Like tremolite, asbestiform actinolite is regulated as asbestos.

Occurrence

Actinolite is commonly found in metamorphic rocks, such as contact aureoles surrounding cooled intrusive igneous rocks. It also occurs as a product of metamorphism of magnesium-rich limestones.

The old mineral name *uralite* is at times applied to an alteration product of primary pyroxene by a mixture composed largely of actinolite. The metamorphosed gabbro or diabase rock bodies, referred to as epidiorite, contain a considerable amount of this *uralitic* alteration.

Some forms of asbestos are formed from fibrous actinolite, the fibres being so small that they can enter the lungs and damage the alveoli.

Gemology

Some forms of actinolite are used as gemstones. One is nephrite, one of the two types of jade (the other being jadeite, a variety of pyroxene).

Another gem variety is the chatoyant form known as *cat s-eye actinolite*. This stone is translucent to opaque, and green to yellowish green colour. This variety has had the misnomer *jade cat s-eye*.

Transparent actinolite is rare and is faceted for gem collectors. Major sources for these forms of actinolite are Taiwan and Canada. Other sources are Madagascar, Tanzania, and the United States.

Acuminite

Acuminite is a rare halide mineral of with chemical formula: $SrAlF_4(OH)\cdot(H_2O)$. Its name comes from the Latin word *acumen*, meaning "spear point". Its Mohs scale rating is 3.5.

Acumenite has only been described from its type locality of the cryolite deposit in Ivigtut, Greenland.

Adamite

Adamite is a zinc arsenate hydroxide mineral, Zn_2AsO_4OH. It is a mineral that typically occurs in the oxidized or weathered zone above zinc ore occurrences. Pure adamite is colourless, but usually it possess yellow colour due to Fe compounds admixture. Tints of green also occur and are connected with copper substitutions in the mineral structure.

Olivenite is a copper arsenate that is isostructural with adamite and there is considerable substitution between zinc and copper resulting in an intermediate called *cuproadamite*. Zincolivenite is a recently discovered mineral being an intermediate mineral with formula $CuZn(AsO_4)(OH)$. Manganese, cobalt, and nickel also substitute in the structure. An analogous zinc phosphate, *tarbuttite*, is known.

Adamite on limonite from the Gold Hill District Tooele County, Utah, USA-Scale at bottom is approx. 2.5 cm.

Occurrence

Adamite occurs as a secondary mineral in the oxidized zone of zinc-and arsenic-bearing hydrothermal mineral deposits. It occurs in association with smithsonite, hemimorphite, scorodite, olivenite, calcite, quartz and iron and manganese oxides.

The yellow to bright lime-green coloured crystals and druze along with its distinctive fluorescence make adamite a favourite among mineral collectors. Found in Mapimi, Mexico; Greece; and California and Utah in the United States.

Adamite was named after the French mineralogist Gilbert-Joseph Adam (1795-1881). It was first described in 1866 for an occurrence at the type locality of Chañarcillo, Copiapó Province, Atacama Region, Chile.

Adamsite-(Y)

Adamsite-(Y) (previously IMA1999-020) $NaY(CO_3)_2 \cdot 6H_2O$ is a mineral of sodium, yttrium, carbon, oxygen, and hydrogen. It is named after Prof. Frank Dawson Adams, a Canadian geologist. Its Mohs scale rating is 3.

Adelite

The rare mineral adelite, is a calcium, magnesium, arsenate with chemical formula $CaMgAsO_4OH$. It forms a solid solution series with the vanadium bearing mineral gottlobite. Various transition metals substitute for magnesium and lead replaces calcium leading to a variety of similar minerals in the *adelite-duftite group*.

Adelite forms variably coloured (blue, green, yellow and grey) crystals in the orthorhombic crystal system. The form is typically massive. It has a Mohs hardness rating of 5 and a specific gravity of 3.73 to 3.79.

It was first described in 1891 from Värmland, Sweden. Its name comes from the Greek word for *indistinct*.

Admontite

Admontite is a magnesium borate mineral with formula $MgB_6O_{10} \cdot 7H_2O$. It is named after Admont, Austria. Its Mohs scale rating is 2 to 3.

Aegirine

Aegirine is a member of the clinopyroxene group of inosilicates. Aegirine is the sodium endmember of the aegirine-augite series. Aegirine has the chemical formula $NaFeSi_2O_6$ in which the iron is present as Fe^{3+}. In the aegirine-augite series the sodium is variably replaced by calcium with iron(II) and magnesium replacing the iron(III) to balance the charge. Aluminium also substitutes for the iron(III). It is also known as *acmite*, which is a fibrous, green-coloured variety.

Aegirine occurs as dark green monoclinic prismatic crystals. It has a glassy luster and perfect cleavage. The Mohs hardness varies from 5 to 6 and the specific gravity is 3.2 to 3.4.

Commonly occurs in alkalic igneous rocks, nepheline syenites, carbonatites and pegmatites. Also in regionally metamorphosed schists, gneisses, and iron formations; in blueschist facies rocks, and from sodium metasomatism in granulites. It may occur as an authigenic mineral in shales and marls. It occurs in association with potassic feldspar, nepheline, riebeckite, arfvedsonite, aenigmatite, astrophyllite, catapleiite, eudialyte, serandite and apophyllite.

Figure 2 : Aegirine

Localities include Mont Saint-Hilaire, Quebec, Canada; Kongsberg, Norway; Narsarssuk, Greenland; Kola Peninsula, Russia; Magnet Cove, Arkansas, USA; Kenya; Scotland and Nigeria.

It was first described in 1835 for an occurrence in Rundemyr, Øvre Eiker, Buskerud, Norway. Aegirine was named after Ægir, the Teutonic god of the sea. A synonym for the mineral is *acmite* in reference to the typical pointed crystals.

Aenigmatite

Aenigmatite, also known as *Cossyrite* after Cossyra, the ancient name of Pantelleria, is a sodium, iron, titanium inosilicate mineral. The chemical formula is $Na_2Fe^{2+}{}_5TiSi_6O_{20}$ and its structure consists of single tetrahedral chains with a repeat unit of four and complex side branches. It forms brown to black triclinic lamellar crystals. It has Mohs hardness of 5.5 to 6 and specific gravity of 3.74 to 3.85. Aenigmatite forms a solid-solution series with wilkinsonite, $Na_2Fe^{2+}{}_4Fe^{3+}{}_2Si_6O_{20}$.

Aenigmatite is primarily found in peralkaline volcanic rocks, pegmatites, and granites as well as silica-poor intrusive rocks. It was first described by August Breithaupt in 1865 for an occurrence in the Ilimaussaq intrusive complex of southwest Greenland.

It was also reported from the Kaidun meteorite, possibly a Mars meteorite, which landed on March 1980 in South Yemen. Other notable studied occurrences include:

- Narsaarsuk and elsewhere in Greenland.
- The Khibiny and Lovozero alkaline massifs on Kola Peninsula, Russia.
- The volcanic island of Pantelleria, Italy.
- In the USA, from Granite Mountain, near Little Rock, Pulaski County, Arkansas, and Santa Rosa, Sonoma County, California.
- In Australia, from Warrumbungle volcano, Nandewar volcano, and the Mount Warning complex, New South Wales; and the Peak Range Province, Queensland.
- In Canada, from Mount Edziza, the Ilgachuz and Rainbow Range shield complexes.
- From Logan Point quarry, Dunedin volcano, New Zealand.

Aerinite

Aerinite ($Ca_4(Al,Fe,Mg)_{10}Si_{12}O_{35}(OH)_{12}CO_3 \cdot 12H_2O$) is a bluish-purple inosilicate mineral. It crystallizes in the monoclinic system and occurs as fibrous masses and coatings.

It has a dark, vitreous luster, a specific gravity of 2.48 and a Mohs hardness of 3.

It is a low-temperature hydrothermal phase occurring in zeolite facies alteration of mafic rocks. Associated minerals include prehnite, scolecite and mesolite.

Its name comes from a Greek root "aerinos," meaning "atmosphere" or "sky".

Aerugite

Aerugite is a rare complex nickel arsenate mineral with a variably reported formula: $Ni_9(AsO_4)_2AsO_6$. It forms green to deep blue-green trigonal crystals. It has a Mohs hardness of 4 and a specific gravity of 5.85 to 5.95.

It was first described in 1858 in either the South Terres mine of Cornwall, England or in Erzgebirge, Saxony, Germany. The origin is disputed. The most common occurrence is as an incrustation on furnace walls in which ores are *roasted*. Its name comes from the Greek word *aerugo* for *copper rust*.

Aeschynite-(Ce)

Aeschynite-(Ce) (or Aschynite, Eschinite, Eschynite) is a rare earth mineral of cerium, calcium, iron, thorium, titanium, niobium, oxygen, and hydrogen with formula: $(Ce,Ca,Fe,Th)(Ti,Nb)_2(O,OH)_6$. Its name comes from the Greek word for "shame". The "-(Ce)" means it has more cerium than the yttrium variety aeschynite-(Y). Its Mohs scale rating is 5-6.

Aeschynite-(Nd)

Aeschynite-(Nd) is a rare earth mineral of neodymium, cerium, calcium, thorium, titanium, niobium, oxygen, and hydrogen with formula: $(Nd,Ce,Ca,Th)(Ti,Nb)_2(O,OH)_6$. Its name comes from the Greek word for "shame". Its Mohs scale rating is 5 to 6.

It was first reported for an occurrence in Bayan Obo Inner Mongolia in 1982. In that rare earth mining deposit it occurs in veins within metamorphosed dolomite and slate. It occurs associated with aegirine, riebeckite, barite, fluorite, albite, phlogopite and magnetite.

Aeschynite-(Y)

Aeschynite-(Y) (or Aeschinite-(Y), Aeschynite-(Yt), Blomstrandine, Priorite) is a rare earth mineral of yttrium, calcium, iron, thorium, titanium, niobium, oxygen, and hydrogen with formula: $(Y,Ca,Fe,Th)(Ti,Nb)_2(O,OH)_6$. Its name comes from the Greek word for "shame". Its Mohs scale rating is 5 to 6.

Abelsonite

Abelsonite (synonymous with nickel porphyrin), is an organic porphyrin mineral first described in 1975 for specimens in oil shale from the Green River Formation in eastern Uintah County, Utah. It is named after Philip Hauge Abelson, an American physicist.

Afghanite

Afghanite, $(Na,Ca,K)_8[Al_6Si_6O_{24}](SO_4,Cl_2,CO_3)_3 \cdot 0.5H_2O$, is a hydrous sodium, calcium, potassium, sulfate, chloride, carbonate alumino-silicate mineral. Afghanite is a feldspathoid of the cancrinite group and typically occurs with sodalite group minerals. It forms blue to colourless, typically massive crystals in the hexagonal crystal system. It has a Mohs hardness of 5.5 to 6 and a specific gravity of 2.55 to 2.65. It has refractive index values of $n\omega$=1.523 and $n\varepsilon$=1.529. It has one direction of perfect cleavage and exhibits conchoidal fracture. It fluoresces a bright orange. It was discovered in 1968 in the Lapis-lazuli Mine, Sar-e-Sang, Badakhshan Province, Afghanistan and takes its name from that country.

It has also been described from localities in Germany, Italy, the Pamir Mountains, Tajikistan, near Lake Baikal in Siberia, New York and Newfoundland. It occurs as veinlets in lazurite crystals in the Afghan location and in altered limestone xenoliths within pumice in Pitigliano, Tuscany, Italy.

Afwillite

Afwillite is a calcium hydroxide nesosilicate mineral with formula $Ca_3(SiO_3OH)_2.2H_2O$. It occurs as glassy, colourless to white prismatic monoclinic crystals. Its Mohs scale hardness is between 3 and 4. It occurs as an alteration mineral in contact metamorphism of limestone.

It was first described in 1925 for an occurrence in the Dutoitspan Mine, Kimberley, South Africa and was named for Alpheus Fuller Williams (1874-1953), a past official of the De Beers diamond company.

Afwillite is typically found in veins of spurrite and it belongs to the nesosilicate sub-class. It is monoclinic, its space group is P2 and its point group is 2.

Formation of Afwillite

It is suggested that afwillite forms in fractured veins of the mineral spurrite. Jennite, afwillite, oyelite and calcite are all minerals that form in layers within spurrite veins. It appears that the afwillite, as well as the calcite, forms from precipitated fluids. The jennite is actually an alteration of the afwillite, but both formed from calcium silicates through hydration.

Laboratory studies determined that afwillite forms at a temperature below 200°C, usually around 100°C. Afwillite and spurrite are formed through contact metamorphism of limestone. Contact metamorphism is caused by the interaction of rock with heat and/or fluids from a nearby crystallizing silicate magma.

Structure and Properties

Afwillite has a complex monoclinic structure, and the silicon tetrahedra in the crystal structure are held together by hydrogen bonds. It has perfect cleavage parallel to its (101) and poor cleavage parallel to its (100) faces. It is biaxial and its 2V angle, the measurement from one optical axis to the other optical axis, is 50 – 56 degrees. When viewed under crossed polarizers in a petrographic microscope, it displays first-order orange colours, giving a maximum birefringence of 0.0167 (determined by using the Michel-Levy chart). Afwillite is optically positive. Additionally, it has a prismatic crystal habit. Under a microscope afwillite looks like wollastonite, which is in the same family as afwillite.

Awfillite is composed of double chains that consist of calcium and silicon polyhedral connected to each other by sharing corners and edges. This causes continuous sheets to form parallel to its miller index [-101] faces. The sheets are bonded together by hydrogen bonds and are all connected by Ca-Si-O bonds (Malik and Jeffery 1976). Each calcium atom is in 6-fold octahedral coordination with the oxygen, and the silicon is in 4-fold tetrahedral coordination around the oxygen. Around each silicon there is one OH group and there are three oxygens that neighbour them. The silicon tetrahedra are arranged so that they share an edge with calcium(1), and silicon(2) shares edges with the calcium(2) and calcium(3) polyhedral. The silicon tetrahedra are held together by the OH group and hydrogen bonding occurs between the hydrogen in the OH and the silicon tetrahedra. Hydrogen bonding is caused because the positive ion, hydrogen, is attracted to negatively charge ions which, in this case, are the silicon tetrahedra.

Occurrence in Concrete

Afwillite is one of the calcium silicates that form when Portland cement sets to form concrete. The cement gets its strength from the hydration of its di-and tri-calcium silicates.

Agardite

Agardite is a mineral group consisting of *agardite-(Ce), agardite-(Nd), agardite-(La), and agardite-(Y)*. They comprise a group of minerals that are hydrous hydrated arsenates of rare earth elements and copper, which contain variable amounts of calcium and sometimes lead. Yttrium, cerium, neodymium, lanthanum as well as trace to minor amounts of other rare earth elements are present in their structure. Agardite-(Y) is probably the most often found representative. The general formula for the group is $(REE,Ca)Cu_6(AsO_4)_3(OH)_6 \cdot 3H_2O$. They form needle-like yellow-green (variably hued) crystals in the hexagonal crystal system.

They were first described in 1970 in the Black Forest, Germany. They were named after Jules Agard, a French geologist.

Agrellite

Agrellite ($NaCa_2Si_4O_{10}F$) is a mineral found in Quebec, Canada. It is also known from few other areas.

Agrellite displays pink fluorescence under both shortwave and longwave ultraviolet light.

It is named in honour of Stuart Olof Agrell.

Agrinierite

Agrinierite ($K_2(Ca,Sr)(UO_2)_3O_3(OH)_2\cdot 5H_2O$) is a mineral often found in the oxidation zone of uranium deposits. It is named for Henry Agrinier, an engineer for the Commissariat à l'Énergie Atomique.

Aguilarite

Aguilarite (Ag_4SeS) is a sulfosalt mineral found in the San Carlos mine in Mexico. Its crystal system is orthorhombic.

Aguilarite

Aguilarite (Ag_4SeS) is a sulfosalt mineral found in the San Carlos mine in Mexico. Its crystal system is orthorhombic.

Ahlfeldite

Ahlfeldite ($(Ni,Co)[SeO_3]\cdot 2H_2O$) is a mineral of secondary origin, found in Bolivia.

Aikinite

Aikinite is a sulfide mineral of lead, copper and bismuth with formula $PbCuBiS_3$. It forms black to grey or reddish brown acicular orthorhombic crystals with a Mohs hardness of 2 to 2.5 and a specific gravity of 6.1 to 6.8. It was originally found in 1843 in the Beryozovskoye deposit, Ural Mountains. It is named after Arthur Aikin (1773–1854), an English geologist.

Ajoite

Ajoite is a hydrated sodium potassium copper aluminium silicate hydroxide mineral. Ajoite has the chemical formula $(Na,K)Cu_7AlSi_9O_{24}(OH)_6\cdot 3(H_2O)$, and minor Mn, Fe and Ca are usually also present in the structure. Ajoite is used as a minor ore of copper.

Discovery

In August 1941 Harry Berman of Harvard University was collecting at Ajo, in Pima County, Arizona, USA. He found specimens of dark blue shattuckite, together with a bluish green mineral which he suspected was a new species. Berman and W T Schaller had planned to collaborate on the investigation of this mineral, together with other known copper silicate minerals, but Berman died in a plane crash in 1944, aged 42, before this study was done.

It was not until 1958 that Schaller, together with Angelina Vlisidis (both of the US Geological Survey) studied the greenish mineral and determined that it was indeed a new species. They named it "ajoite" (pronounced ah-hoe-ite) after the place where it had been found.

Twenty three years later, in 1981, George Chao from Carleton University in Ottawa, Canada, studied the mineral again, and showed that ajoite was triclinic, and not monoclinic as had been thought previously. His studies also resulted in a re-definition of the chemical formula.

Crystal Structure

The 1981 investigations of ajoite were done on a sample from the New Cornelia Mine, Ajo, with tiny crystals averaging 0.01×0.1×0.4 mm. The mineral was shown to belong to the triclinic system, but it was not clear whether the crystal class was 1 or 1. The triclinic class 1 has the lowest possible symmetry with no symmetry elements at all, and 1 has just a centre of symmetry, but no axes of symmetry or mirror planes. The space group is either P1 or P1.

The unit cell parameters are:

$a = 13.637$ Å,
$b = 14.507$ Å,
$c = 13.620$ Å,
$\alpha = 107.16°$,
$\beta = 105.45°$ and
$\gamma = 110.57°$.

Ajoite occurs as sprays of bladed prismatic crystals, commonly fibrous, elongated along the c crystal axis and flattened sideways. The dominant form is {010}, parallel to the plane containing the a and c axes. The forms {110}, cutting both the a and b axes, and {100}, parallel to the plane containing the b and c axes, are much less prominent but are always present. The termination on c may be either {001} or {203} or both.

Ajoite is frequently found as inclusions in quartz, much sought after by mineral collectors.

Optical Properties

Ajoite is translucent and bluish green in colour, with a pale greenish white streak and a vitreous luster. The optical class is biaxial (+) with refractive indices $N_x = 1.550$, $N_y = 1.583$ and $N_z = 1.641$. These values are similar to the refractive index of ordinary window glass, at 1.5. In white light the mineral is distinctly pleochroic, being very light bluish green along the X direction and brilliant bluish green along the Y and Z directions.

Physical Properties

The cleavage is perfect parallel to the plane containing the a and c axes, and the hardness is 3.5, between that of calcite and fluorite. It is a little denser than quartz, with specific gravity 2.96.

Ajoite is readily decomposed by hydrochloric acid, HCl, and by nitric acid, HNO_3, but it does not react with ammonium hydroxide, NH_4OH. It has been reported as having barely detectable radioactivity.

Environment

Ajoite is a secondary mineral that forms from the oxidation of other secondary copper minerals in copper-rich base metal deposits in massive fracture coatings, in vein fillings, and in vugs. It may form from shattuckite and also it may be replaced by shattuckite.

At the type locality it is associated with shattuckite, conichalcite, quartz, muscovite and pyrite.

Occurrence

Ajoite is named after its type locality, the New Cornelia Mine in the Ajo District of Pima County, Arizona. Type material is conserved at the National Museum of Natural History, Washington DC, USA, reference number 113220.

Other localities include Wickenburg and Maricopa County in Arizona, within the United States, and the Messina District in South Africa. Quartz specimens from the defunct Messina Mines on the border between Zimbabwe and South Africa are well known for their inclusions of blue copper silicate minerals such as shattuckite, papagoite and ajoite, but ajoite from American localities does not occur like this.

Akaganéite

Akaganéite is an iron(III) oxide-hydroxide/chloride mineral with formula: $Fe^{3+}O(OH,Cl)$ e.g; β-FeO(OH). It is formed by the weathering of pyrrhotite ($Fe_{1-x}S$). Nickel also has been reported in the structure. It has a monoclinic crystal structure, a metallic luster and a brownish yellow streak.

It is named after the Akagane mine in Iwate, Japan, in which it was discovered. It has also been found in widely dispersed locations around the world and in rocks from the Moon that were brought back during the Apollo Project. The occurrences in meteorites and the lunar sample are thought to have been produced by interaction with Earth's atmosphere.

Akatoreite

Akatoreite ($(Mn^{2+}, Fe^{2+})_9Al_2[(OH)_3 | HSi_4O_{13}]$) is a mineral found in New Zealand.

Akdalaite

Akdalaite is a very rare mineral found in Kazakhstan and has the formula $5Al_2O_3 \cdot H_2O$. It was formerly believed to be $4Al_2O_3 \cdot H_2O$. It is therefore the same as tohdite an artificailly produced phase. Studies on the crystal structure and spectra indicate that this is an aluminium oxide hydroxide.

Akermanite

Akermanite ($Ca_2Mg[Si_2O_7]$) is a melilite mineral of the sorosilicate group, containing calcium, magnesium, silicon, and oxygen. It is a product of contact metamorphism of siliceous limestones and dolostones, and rocks of sanidinite facies.

Sanidinite facies represent the highest conditions of temperature of contact metamorphism and are characterized by the absence of hydrous minerals.

It has a density of 2.944 g/cm³. Åkermanite ranks a 5 or 6 on the Mohs scale of mineral hardness, and can be found grey, green, brown, or colourless. It has a white streak and a vitreous or resinous luster. It has a tetragonal crystal system and a good, or distinct, cleavage. It is the end member in a solid solution series beginning with gehlenite ($Ca_2Al[AlSiO_2]$).

The mineral is named for Anders Richard Åkerman, a Swedish metallurgist. It has been found at Monte Somma and Vesuvius, and Monte Cavalluccio near Rome. It was "grandfathered" in as a species of mineral because it was described prior to 1959, before the founding of the International Mineralogical Association.

Agate

Agate is a microcrystalline variety of silica, chiefly chalcedony, characterised by its fineness of grain and brightness of colour. Although agates may be found in various kinds of rock, they are classically associated with volcanic rocks and can be common in certain metamorphic rocks.

Etymology and History

The stone was given its name by Theophrastus, a Greek philosopher and naturalist, who discovered the stone along the shore line of the river Achates sometime between the 4th and 3rd centuries BC. Colourful agates and other chalcedonies were obtained over 3,000 years ago from the Achates River, now called Dirillo, in Sicily.

Ancient use

Agate is one of the most common materials used in the art of hardstone carving, and has been recovered at a number of ancient sites, indicating its widespread use in the ancient world; for example, archaeological recovery at the Knossos site on Crete illustrates its role in Bronze Age Minoan culture.

Formation and Characteristics

Most agates occur as nodules in volcanic rocks or ancient lavas where they represent cavities originally produced by the disengagement of volatiles in the molten mass which were then filled, wholly or partially, by siliceous matter deposited in regular layers upon the walls. Such agates, when cut transversely, exhibit a succession of parallel lines, often of extreme tenuity, giving a banded appearance to the section. Such stones are known as banded agate, riband agate and striped agate.

In the formation of an ordinary agate, it is probable that waters containing silica in solution—derived, perhaps, from the decomposition of some of the silicates in the lava itself—percolated through the rock and deposited a siliceous coating on the interior of the vapour-vesicles. Variations in the character of the solution or in the conditions of deposition may cause a corresponding variation in the successive layers, so that bands of chalcedony often alternate with layers of crystalline quartz.

Several vapour-vesicles may unite while the rock is still viscous, and thus form a large cavity which may become the home of an agate of exceptional size; thus a Brazilian geode lined with amethyst and weighing 35 tons was exhibited at the Düsseldorf Exhibition of 1902. Perhaps the most comprehensive review of agate chemistry.

The first deposit on the wall of a cavity, forming the "skin" of the agate, is generally a dark greenish mineral substance, like celadonite, delessite or "green earth", which are rich in iron probably derived from the decomposition of the augite in the enclosing volcanic rock. This green silicate may give rise by alteration to a brown iron oxide (limonite), producing a rusty appearance on the outside of the agate-nodule.

The outer surface of an agate, freed from its matrix, is often pitted and rough, apparently in consequence of the removal of the original coating. The first layer spread over the wall of the cavity has been called the "priming", and upon this base zeolitic minerals may be deposited.

Many agates are hollow, since deposition has not proceeded far enough to fill the cavity, and in such cases the last deposit commonly consists of quartz, often amethyst, having the apices of the crystals directed towards the free space so as to form a crystal-lined cavity, or geode.

On the disintegration of the matrix in which the agates are embedded, they are set free. The agates are extremely resistant to weathering and remain as nodules in the soil or are deposited as gravel in streams and shorelines.

Types of Agate

Figure 3 Agatized Coral

Figure 4: "Turritella agate" (Elimia tenera) from Green River Formation, Wyoming

A Mexican agate, showing only a single eye, has received the name of *cyclops agate.* Included matter of a green, golden, red, black or other colour or combinations embedded in the chalcedony and disposed in filaments and other forms suggestive of vegetable growth, gives rise to dendritic or moss agate. *Dendritic* agates have fern like patterns in them formed due to the presence of manganese and iron oxides. Other types of included matter deposited during agate-building include sagenitic growths (radial mineral crystals) and chunks of entrapped detritus (such as sand, ash, or mud). Occasionally agate fills a void left by decomposed vegetative material such as a tree limb or root and is called limb cast agate due to its appearance.

Turritella agate is formed from silicified fossil *Elimia tenera* (erroneously considered *Turritella*) shells. *E. tenera* are spiral marine gastropods having elongated, spiral shells composed of many whorls. Similarly, coral, petrified wood and other organic remains or porous rocks can also become agatized. Agatized coral is often referred to as Petoskey stone or agate.

Greek agate is a name given to pale white to tan coloured agate found in Sicily back to 400 B.C. The Greeks used it for making jewellery and beads. Even though the stone had been around centuries and was known to both the Sumerians and the Egyptians, both who used the gem for decoration and for playing important parts in their religious ceremonies, any agate of this colour from Sicily, once an ancient Greek colony, is called Greek agate.

Another type of agate is Brazilian agate, which is found as sizable geodes of layered nodules. These occur in brownish tones interlayered with white and gray. Quartz forms within these nodules, creating a striking specimen when cut opposite the layered growth axis. It is often dyed in various colours for ornamental purposes.

Certain stones, when examined in thin sections by transmitted light, show a diffraction spectrum due to the extreme delicacy of the successive bands, whence they are termed *rainbow agates*. Often agate coexists with layers or masses of opal, jasper or crystalline quartz due to ambient variations during the formation process.

Other forms of agate include Lake Superior agate, carnelian agate (usually exhibiting reddish hues), Botswana agate, Ellensburg blue agate, blue lace agate, plume agates, tube agate (with visible flow channels), fortification agate (which exhibit little or no layered structure), *fire agate* (which seems to glow internally like an opal) and Mexican crazy-lace agate (which exhibits an often brightly coloured, complex banded pattern) also called Rodeo Agate and Rosetta Stone depending on who owned the mine at the time.

Uses in Industry and Art

Industry uses agates chiefly to make ornaments such as pins, brooches, paper knives, inkstands, marbles and seals. Because of its hardness and ability to resist acids, agate is used to make mortars and pestles to crush and mix chemicals. Because of the high polish possible with agate it has been used for centuries for leather burnishing tools. Idar-Oberstein was one of the centres which made use of agate on an industrial scale. Where in the beginning locally found agates were used to make all types of objects for the European market, this became a globalized business around the turn of the 20th century: Idar-Oberstein imported large quantities of agate from Brazil, as ship's ballast. Making use of a variety of proprietary chemical processes, they produced coloured beads that were sold around the globe. Agates have long been used in arts and crafts. The sanctuary of a Presbyterian church in Yachats, Oregon, has six windows with panes made of agates collected from the local beaches.

Alabaster

Alabaster is a name applied to varieties of two distinct minerals, when used as a material: gypsum (a hydrous sulfate of calcium) and calcite (a carbonate of calcium). The former is the alabaster of the present day; generally, the latter is the alabaster of the ancients. Both are easy to work, with an attractive appearance, and have been used for making a variety of artworks and objects, especially small carvings.

The two kinds are distinguished from one another readily, because of differences in their relative hardness. The gypsum kind is so soft as to be readily scratched with a fingernail (Mohs hardness 1.5 to 2), while the calcite kind is too hard to be scratched in this way (Mohs hardness 3), although it does yield readily to a knife. Moreover, the calcite alabaster, being a carbonate, effervesces upon being touched with hydrochloric acid, whereas the gypsum alabaster, when thus treated, remains practically unaffected.

Due to the characteristic colour of white alabaster, the term has entered the vernacular as a metonym for white things, particularly "alabaster skin", which means very light and quite transparent, and no doubt derives from the use of alabaster for tomb effigies.

Etymology

The origin of the word, alabaster, is in Middle English, through Old French *alabastre*, in turn derived from the Latin *alabaster* and that from Greek *alabastros* or *alabastos*. The latter was a term used to identify a vase made of alabaster.

This name may derive further from the Ancient Egyptian word *a-labaste*, which refers to vessels of the Egyptian goddess Bast. She was represented as a lioness and frequently depicted as such in figures placed atop these alabaster vessels. Other suggestions include derivation from the town of Alabastron in Egypt.

Types

Calcite Alabaster

This substance, the "alabaster" of the Ancient Egyptians and Bible, often is termed *Oriental alabaster*, since the early examples came from the Far East. The Greek name *alabastrites* is said to be derived from the town of Alabastron, in Egypt, where the stone was quarried. The locality probably owed its name to the mineral; the origin of the mineral name is obscure.

The "Oriental" alabaster was highly esteemed for making small perfume bottles or ointment vases called alabastra, the vessel name has been suggested as a possible source of the mineral name. In Egypt craftsmen used alabaster for canopic jars and various other sacred and sepulchral objects. A sarcophagus, sculptured in a single block of translucent calcite alabaster from Alabastron, is in the Sir John Soane's Museum, London. This was discovered by Giovanni Belzoni in 1817 in the tomb of Seti I near Thebes. It was purchased by Sir

John Soane. When cut in thin sheets, alabaster is translucent enough to be used for small windows. It was used for this purpose in medieval churches, especially in Italy. Large alabaster sheets are used extensively in the contemporary Cathedral of Our Lady of the Angels, which was dedicated in 2002 by the Los Angeles, California Archdiocese. The cathedral incorporates special cooling to prevent the panes from overheating and turning opaque.

Calcite alabaster is found as either a stalagmitic deposit, from the floor and walls of limestone caverns, or as a kind of travertine, similarly deposited in springs of calcareous water. Its deposition in successive layers gives rise to the banded appearance that the marble often shows on cross-section, whence it is known as onyx-marble or alabaster-onyx, or sometimes simply as onyx. The latter term should be restricted to siliceous minerals, however.

Egyptian alabaster has been worked extensively near Suez and Assiut. Many ancient quarries are found in the hills overlooking the plain of Tell el Amarna.

Algerian onyx-marble has been quarried largely in the province of Oran. In Mexico, there are famous deposits of a delicate green variety at La Pedrara, in the district of Tecali, near Puebla. Onyx-marble occurs also in the district of Tehuacán and at several localities in California, Arizona, Utah, Colorado, and Virginia.

Gypsum Alabaster

When the term "alabaster" is used without any qualification, it invariably means a fine-grained variety of gypsum. This mineral, or alabaster proper, occurs in England. The early use of alabaster for vessels dedicated for use in the cult of the deity Bast in the culture of the Ancient Egyptians is well documented, however, thousands of gypsum alabaster artifacts dating to the late 4th millennium BC also have been found in Tell Brak (present day Nagar), in Syria. In Mesopotamia, a gypsum alabaster sculpture, believed to represent the deity, Abu, dates to the first half of the 3rd millennium BC.

Gypsum alabaster is a common mineral, which occurs in England in the Keuper marls of the Midlands, especially at Chellaston in Derbyshire, at Fauld in Staffordshire, and near Newark in Nottinghamshire. Deposits at all of these localities have been worked extensively. In the fifteenth century its carving into small statues and sets of relief panels for altarpieces was a valuable local industry in Nottingham, as well as a major English export. These were usually painted, or partly painted. It was also used for the effigies, often life size, on tomb monuments, as the typical recumbent position suited the material's lack of strength, and it was cheaper and easier to work than good marble. Beside examples of these carvings still in Britain (especially at the Nottingham Castle Museum, British Museum, and Victoria and Albert Museum), trade in mineral alabaster (rather than just the antiques trade) has scattered examples in the material that may be found as far afield as the Musée de Cluny, Spain, and Scandinavia.

Alabaster also is found, although in smaller quantity, at Watchet in Somerset, near Penarth in Glamorganshire, and elsewhere. In Cumbria it occurs largely in the New Red rocks, but at a lower geological horizon. The alabaster of Nottinghamshire and Derbyshire is found in thick nodular beds or "floors" in spheroidal masses known as "balls" or "bowls" and in smaller lenticular masses termed "cakes." At Chellaston, where the local alabaster is known as "Patrick," it has been worked into ornaments under the name of "Derbyshire spar"—a term more properly applied to fluorspar.

In Europe, the centre of the alabaster trade today is Florence, Italy. Tuscan alabaster occurs in nodular masses embedded in limestone, interstratified with marls of Miocene and Pliocene age. The mineral is worked largely by means of underground galleries, in the district of Volterra. Several varieties are recognized—veined, spotted, clouded, agatiform, and others. The finest kind, obtained principally from Castellina, is sent to Florence for figure-sculpture, while the common kinds are carved locally, into vases, lights, and various ornamental objects. These items are objects of extensive trade, especially in Florence, Pisa, and Livorno. Spain and many other countries have deposits which have been worked.

Black Alabaster

Alabaster Caverns State Park, near Freedom, Oklahoma is home to a natural gypsum cave in which much of the gypsum is in the form of alabaster. There are several types of alabaster found at the site, including pink, white, and the rare black alabaster.

Uses

The finer kinds of alabaster are employed largely as an ornamental stone, especially for ecclesiastical decoration and for the rails of staircases and halls. Its softness enables it to be carved readily into elaborate forms, but its solubility in water renders it unsuitable for outdoor work. If alabaster with a smooth, polished surface is washed with washing-up liquid (dishwashing liquid), it will become rough, dull and whiter, losing most of its translucency and lustre.

The purest alabaster is a snow-white material of fine tiniforni grain, but it often is associated with an oxide of iron, which produces brown clouding and veining in the stone. The coarser varieties of alabaster are converted by calcination into plaster of Paris, whence they sometimes are known as "plaster stone."

In order to diminish the translucency of the alabaster and to produce an opacity suggestive of true marble, the statues are immersed in a bath of water and heated gradually—nearly to the boiling-point—an operation requiring great care, for if the temperature is not regulated carefully, the stone acquires a dead-white, chalky appearance. The effect of heating appears to be a partial dehydration of the gypsum. If properly treated, it very closely resembles true marble and is known as marmo di Castellina.

Sulphate of lime (gypsum) also was used by the ancients. It was employed, for instance, in Assyrian sculpture. So some of the ancient alabaster is identical to the modern stone.

Alabaster may be stained to disguise it, by being heated in various pigmentary solutions. In this way a very misleading imitation of coral that is called alabaster coral is produced.

Chrysoberyl

The mineral or gemstone chrysoberyl is an aluminate of beryllium with the formula $BeAl_2O_4$. The name chrysoberyl is derived from the Greek words *chrysos* and *beryllos*, meaning "a gold-white spar". Despite the similarity of their names, chrysoberyl and beryl are two completely different gemstones. Chrysoberyl is the third-hardest frequently encountered natural gemstone and lies at 8.5 on the hardness scale, between corundum (9) and topaz (8).

An interesting feature of its crystals are the cyclic twins called *trillings*. These twinned crystals have a hexagonal appearance, but are the result of a triplet of twins with each "twin" oriented at 120° to its neighbours and taking up 120° of the cyclic trilling. If only two of the three possible twin orientations are present, a "V"-shaped twin results.

Ordinary chrysoberyl is yellowish-green and transparent to translucent. When the mineral exhibits good pale green to yellow colour and is transparent, then it is used as a gemstone. There are three main varieties of chrysoberyl: ordinary yellow-to-green chrysoberyl, cat's eye or cymophane, and alexandrite. Yellow-green chrysoberyl was referred to as "chrysolite" during the Victorian and Edwardian eras, which caused confusion since that name has also been used for the mineral olivine ("peridot" as a gemstone); that name is no longer used in the gemological nomenclature.

Alexandrite, a strongly pleochroic (trichroic) gem, will exhibit emerald green, red and orange-yellow colours depending on viewing direction in partially polarised light. However, its most distinctive property is that it also changes colour in artificial (tungsten/halogen) light compared to daylight. The colour change from red to green is due to strong absorption of light in a narrow yellow portion of the spectrum, while allowing large bands of blue-greener and red wavelengths to be transmitted. Which of these prevails to give the perceived hue depends on the spectral balance of the illumination. Typically, alexandrite has an emerald-green colour in daylight (relatively blue illumination of high colour temperature) but exhibits a raspberry-red colour in incandescent light (relatively yellow illumination).

Cymophane is popularly known as "cat's eye". This variety exhibits pleasing chatoyancy or opalescence that reminds one of an eye of a cat. When cut to produce a cabochon, the mineral forms a light-green specimen with a silky band of light extending across the surface of the stone.

Occurrence

Chrysoberyl forms as a result of pegmatitic processes. Melting in the Earth's crust produces relatively low-density molten magma which can rise upwards towards the surface. As the main magma body cools, water originally present in low concentrations became more concentrated in the molten rock because it could not be incorporated into the crystallization of solid minerals. The remnant magma thus becomes richer in water, and also in rare elements that similarly do not fit in the crystal structures of major rock-forming minerals. The water extends the temperature range downwards before the magma sets solid completely, allowing concentration of rare elements to proceed so far that they produce their own distinctive minerals. The resulting rock, igneous in appearance but formed at a low temperature from a water-rich melt, with large crystals of the common minerals such as quartz and feldspar, but also with elevated concentrations of rare elements such as beryllium, lithium, or niobium, often forming their own minerals, is called a pegmatite. The high water content of the magma made it possible for the crystals to grow quickly, so pegmatite crystals are often quite large, which increases the likelihood of gem specimens forming.

Chrsoberyl can also grow in the country rocks near to pegmatites, when Be-and Al-rich fluids from the pegmatite react with surrounding minerals. Hence, it can be found in mica schists and in contact with metamorphic deposits of dolomitic marble. Because it is a hard, dense mineral that is resistant to chemical alteration, it can be weathered out of rocks and deposited in river sands and gravels in alluvial deposits with other gem minerals such as diamond, corundum, topaz, spinel, garnet, and tourmaline. When found in such placers, it will have rounded edges instead of sharp, wedge-shape forms. Much of the chrysoberyl mined in Brazil and Sri Lanka is recovered from placers as the host rocks have been intensely weathered and eroded.

If the pegmatite fluid is rich in beryllium, crystals of beryl or chrysoberyl could form. Beryl has a high ratio of beryllium to aluminium, while the opposite is true for chrysoberyl. Both are stable with the common mineral quartz. For alexandrite to form, some chromium would also have had to be present. However, beryllium and chromium do not tend to occur in the same types of rock. Chromium is commonest in mafic and ultramafic rocks in which beryllium is extremely rare. Beryllium becomes concentrated in felsic pegmatites in which chromium is almost absent. Therefore, the only situation where an alexandrite can grow is when Be-rich pegmatitic fluids react with Cr-rich country rock. This unusual requirement explains the rarity of this chrysoberyl variety.

Chrysoberyl

Chrysoberyl was discovered in 1789 and described and named by Abraham Gottlob Werner, in 1790. Werner worked at the Freiberg Mining Academy from 1790–1793 and was well known as one of the most outstanding geologists of his time. He is best known today as the loser in the battle of the Neptunists and Vulcanists that raged in the 1780s.

Chrysoberyl is normally yellow, yellow-green, or brownish with its colour being caused by the presence of iron. Spectroscopic analysis will usually reveal a strong band where the violet takes over from the blue. As the colour darkens from bright yellowish-green to golden-yellow to brown, this band increases in strength. When the stone has a strong colour, two additional bands can be seen in the green-blue. The most common inclusions are liquid-filled cavities containing three-phase inclusions. Stepped twin planes may be apparent in some cases. Some very rare minty bluish-green chrysoberyls from Tanzania owe their colour to the presence of vanadium.

Despite the similarity of their names, chrysoberyl and beryl are two completely different gemstones. Members of the beryl group include emerald, aquamarine, and morganite while members of the chrysoberyl group include chrysoberyl, cymophane (cat's eye), and alexandrite. Beryl is a silicate and chrysoberyl is an oxide and although both beryl and chrysoberyl contain beryllium, they are separate gemstone species unrelated in any other way. Because of the confusion between chrysoberyl and beryl, chrysoberyl is relatively unknown in its own right and the alexandrite variety is much more widely recognized. The only well-known natural gemstones harder than chrysoberyl are corundum and diamond.

Alexandrite

The alexandrite variety displays a colour change (*alexandrite effect*) dependent upon the nature of ambient lighting. This colour shift is independent of any change of hue with viewing direction through the crystal that would arise from pleochroism. Both these different properties are frequently referred to as "colour change", however.

Alexandrite results from small scale replacement of aluminium by chromium ions in the crystal structure, which causes intense absorption of light over a narrow range of wavelengths in the yellow region of the spectrum. Alexandrite from the Ural Mountains in Russia is green by daylight and red by incandescent light. Other varieties of alexandrite may be yellowish or pink in daylight and a columbine or raspberry red by incandescent light.

According to a popular but controversial story, alexandrite was discovered by the Finnish mineralogist Nils Gustaf Nordenskiöld, (1792–1866) and named alexandrite in honour of the future Tsar Alexander II of Russia. Nordenskiöld's initial discovery occurred as a result of an examination of a newly found mineral sample he had received from Perovskii, which he identified as emerald at first. The first emerald mine had been opened in 1831.

Alexandrite up to 5 carats (1,000 mg) and larger were traditionally thought to be found only in the Ural Mountains. Other deposits are located in India (Andhra Pradesh), Madagascar, and Sri Lanka. Alexandrite in sizes over 3 carats are very rare.

Some gemstones described as lab-grown (synthetic) alexandrite are actually corundum laced with trace elements (e.g., vanadium) or colour-change spinel and are not actually chrysoberyl. As a result, they would be more accurately described as simulated alexandrite rather than synthetic but are often called Czochralski alexandrite after the process that grows the crystals.

Cymophane

Translucent yellowish chatoyant chrysoberyl is called cymophane or *cat s eye*. Cymophane has its derivation also from the Greek words meaning 'wave' and 'appearance', in reference to the chatoyancy sometimes exhibited. In this variety, microscopic tubelike cavities or needlelike inclusions of rutile occur in an orientation parallel to the c-axis producing a chatoyant effect visible as a single ray of light passing across the crystal. This effect is best seen in gemstones cut in cabochon form perpendicular to the c-axis. The colour in yellow chrysoberyl is due to Fe^{3+} impurities.

Although other minerals such as tourmaline, scapolite, corundum, spinel and quartz can form "cat's eye" stones similar in appearance to cymophane, the jewellery industry designates these stones as "quartz cat's eyes", or "ruby cat's eyes" and only chrysoberyl can be referred to as "cat's eye" with no other designation.

Gems lacking the silky inclusions required to produce the cat's eye effect are usually faceted. An alexandrite cat's eye is a chrysoberyl cat's eye that changes colour. "Milk and honey" is a term commonly used to describe the colour of the best cat's eyes. The effect refers to the sharp milky ray of white light normally crossing the cabochon as a centre line along its length and overlying the honey coloured background. The honey colour is considered to be top-grade by many gemologists but the lemon yellow colours are also popular and attractive. Cat's eye material is found as a small percentage of the overall chrysoberyl production wherever chrysoberyl is found.

Cat's eye really became popular by the end of the 19th century when the Duke of Connaught gave a ring with a cat's eye as an engagement token; this was sufficient to make the stone more popular and increase its value greatly. Until that time, cat's eye had predominantly been present in gem and mineral collections. The increased demand in turn created an intensified search for it in Sri Lanka. Early 20th century prices could go up as high as $8000 for a cut stone.

Amber

Amber is fossilized tree resin (not sap), which has been appreciated for its colour and natural beauty since Neolithic times. Amber is used as an ingredient in perfumes, as a healing agent in folk medicine, and as jewellery. There are five classes of amber, defined on the basis of their chemical constituents. Because it originates as a soft, sticky tree resin, amber sometimes contains animal and plant material as inclusions. Amber occurring in coal seams is also called resinite, and the term ambrite is applied to that found specifically within New Zealand coal seams.

History and Etymology

The English word *amber* derives from the Arabic *anbar*, via Medieval Latin *ambar* and Old French *ambre*. The word originally referred to a precious oil derived from the Sperm whale (now called ambergris). The sense was extended to fossil resin circa 1400, and this became the main sense as the use of ambergris waned. The two substances were confused because they both were found washed up on beaches. Ambergris is less dense than water and floats; whereas amber is less dense than stone, but too dense to float. The word "ambar" was brought to Europe by the Crusaders. In French "ambre gris" was then distinguished from "ambre jaune": ambre gris (gray amber) was ambergris; ambre jaune (yellow amber) was the fossil resin we now call amber.

Amber is discussed by Theophrastus, possibly the first historical mention of the material, in the 4th century BC. The Greek name for amber was *electron* and was connected to the Sun God, one of whose titles was *Elector* or the *Awakener*. The modern terms "electricity" and "electron" derive from the Greek word for amber and come from William Gilbert's research showing that amber could attract other substances. The word "electron" was coined in 1891 by the Irish physicist George Stoney whilst analysing elementary charges for the first time.

The presence of insects in amber was noticed by Pliny the Elder in his Naturalis Historia, and led him to theorize correctly that, at some point, amber had to be in a liquid state to cover the bodies of insects. Hence he gave it the expressive name of *succinum* or *gum-stone*, a name that is still in use today to describe succinic acid as well as *succinite*, a term given to a particular type of amber by James Dwight Dana.

Heating amber will soften it and eventually it will burn, which is why in Germanic languages the word for amber is a literal translation of *burn-Stone*. Heated above 200°C, amber suffers decomposition, yielding an "oil of amber", and leaving a black residue which is known as "amber colophony", or "amber pitch"; when dissolved in oil of turpentine or in linseed oil this forms "amber varnish" or "amber lac".

Amber from the Baltic Sea has been extensively traded since antiquity and in the main land, from where amber was traded 2000 years ago, the natives called it *glaes*.

The Baltic Lithuanian term for amber is *Gintaras* and Latvian *Dzintars*. They and the Slavic *jantar* are thought to originate from Phoenician *jainitar* (sea-resin). However, while most Slavic languages, such as Russian and Czech, retain the old Slavic word, in the Polish language, despite still being correct, *jantar* is used very rarely (even considered archaic) and was replaced by the word *bursztyn* deriving from the German analogue.

Composition and Formation

Amber is heterogeneous in composition, but consists of several resinous bodies more or less soluble in alcohol, ether and chloroform, associated with an insoluble bituminous substance. Amber is a macromolecule by free radical polymerization of several precursors in the labdane family, e.g. communic acid, cummunol, and biformene. These labdanes are diterpenes ($C_{20}H_{32}$) and trienes, equipping the organic skeleton with three alkene groups for polymerization. As amber matures over the years, more polymerization takes place as well as isomerization reactions, crosslinking and cyclization.

The average composition of amber leads to the general formula $C_{10}H_{16}O$.

Formation

Molecular polymerization, resulting from high pressures and temperatures produced by overlying sediment, transforms the resin first into copal. Sustained heat and pressure drives off terpenes and results in the formation of amber.

Botanical Origin

Fossil resins from Europe fall into two categories, the famous Baltic ambers and another that resembles the Agathis group. Fossil resins from the Americas and Africa are closely related to the modern genus Hymenaea, while Baltic ambers are thought to be fossil resins from *Sciadopityaceae* family plants that used to live in north Europe.

Inclusions

The abnormal development of resin has been called *succinosis*. Impurities are quite often present, especially when the resin dropped on to the ground, so that the material may be useless except for varnish-making, whence the impure amber is called *firniss*. Enclosures of pyrites may give a bluish colour to amber. The so-called *black amber* is only a kind of jet. *Bony amber* owes its cloudy opacity to minute bubbles in the interior of the resin.

In darkly clouded and even opaque amber, inclusions can be imaged using high-energy, high-contrast, high-resolution X-rays.

Extraction and Processing

Distribution and Mining

Amber is globally distributed, mainly in rocks of Cretaceous age or younger. Historically, the coast around Königsberg in Prussia was the world's leading source of amber. About 90% of the world's extractable amber is still located in the Kaliningrad Oblast of Russia on the Baltic Sea (which was previously Königsberg in Prussia, before World War II).

Pieces of amber torn from the seafloor are cast up by the waves, and collected by hand, dredging, or diving. Elsewhere, amber is mined, both in open works and underground galleries. Then nodules of *blue earth* have to be removed and an opaque crust must be cleaned off, which can be done in revolving barrels containing sand and water. Erosion removes this crust from sea-worn amber.

Dominican amber, especially Dominican blue amber, is mined through bell pitting, which is dangerous due to the risk of tunnel collapse.

Treatment

The Vienna amber factories, which use pale amber to manufacture pipes and other smoking tools, turn it on a lathe and polish it with whitening and water or with rotten stone and oil. The final lustre is given by friction with flannel. When gradually heated in an oil-bath, amber becomes soft and flexible.

Two pieces of amber may be united by smearing the surfaces with linseed oil, heating them, and then pressing them together while hot. Cloudy amber may be clarified in an oil-bath, as the oil fills the numerous pores to which the turbidity is due. Small fragments, formerly thrown away or used only for varnish, are now used on a large scale in the formation of "amberoid" or "pressed amber". The pieces are carefully heated with exclusion of air and then compressed into a uniform mass by intense hydraulic pressure; the softened amber being forced through holes in a metal plate.

The product is extensively used for the production of cheap jewellery and articles for smoking. This pressed amber yields brilliant interference colours in polarized light. Amber has often been imitated by other resins like copal and kauri, as well as by celluloid and even glass. Baltic amber is sometimes coloured artificially, but also called "true amber".

Appearance

Amber occurs in a range of different colours. As well as the usual yellow-orange-brown that is associated with the colour "amber", amber itself can range from a whitish colour through a pale lemon yellow, to brown and almost black. Other more uncommon colours include red amber (sometimes known as "cherry amber"), green amber, and even blue amber, which is rare and highly sought after.

Much of the most highly-prized amber is transparent, in contrast to the very common cloudy amber and opaque amber. Opaque amber contains numerous minute bubbles. This kind of amber is known as "bony amber".

Although all Dominican amber is fluorescent, the rarest Dominican amber is blue amber. It turns blue in natural sunlight and any other partially or wholly ultraviolet light source. In long-wave UV light it has a very strong reflection, almost white. Only about 100 kg is found per year, which makes it valuable and expensive.

Sometimes amber retains the form of drops and stalactites, just as it exuded from the ducts and receptacles of the injured trees. It is thought that, in addition to exuding onto the surface of the tree, amber resin also originally flowed into hollow cavities or cracks within trees, thereby leading to the development of large lumps of amber of irregular form.

Classification

Amber can be classified into several forms. Most fundamentally, there are two types of plant resin with the potential for fossilization. Terpenoids, produced by conifers and angiosperms, consist of ring structures formed of isoprene (C_5H_8) units. Phenolic resins are today only produced by angiosperms, and tend to serve functional uses. The extinct medullosans produced a third type of resin, which is often found as amber within their veins.

The composition of resins is highly variable; each species produces a unique blend of chemicals which can be identified by the use of pyrolysis–gas chromatography–mass spectrometry. The overall chemical and structural composition is used to divide ambers into five classes. There is also a separate classifications of amber gemstones, according to the way of production.

Class I

This class is by far the most abundant. It comprises labdatriene carboxylic acids such as communic or ozic acids. It is further split into three sub-classes. Classes Ia and Ib utilise regular labdanoid diterpenes (e.g. communic acid, communol, biformenes), whilst Ic uses *enantio* labdanoids (ozic acid, ozol, *enantio* biformenes).

Ia

Includes *Succinite* (= 'normal' Baltic amber) and *Glessite*. Have a communic acid base. They also include much succinic acid.

Baltic amber yields on dry distillation succinic acid, the proportion varying from about 3% to 8%, and being greatest in the pale opaque or *bony* varieties.

The aromatic and irritating fumes emitted by burning amber are mainly due to this acid. Baltic amber is distinguished by its yield of succinic acid, hence the name *succinite*. Succinite has a hardness between 2 and 3, which is rather greater than that of many other fossil resins. Its specific gravity varies from 1.05 to 1.10. It can be distinguished from other ambers via IR spectroscopy due to a specific carbonyl absorption peak. IR spectroscopy can detect the relative age of an amber sample. Succinic acid may not be an original component of amber, but rather a degradation product of abietic acid.

Ib

Like class Ia ambers, these are based on communic acid; however, they lack succinic acid.

Ic

This class is mainly based on *enantio*-labdatrienonic acids, such as ozic and zanzibaric acids. Its most familiar representative is Dominican amber.

Dominican amber differentiates itself from Baltic amber by being mostly transparent and often containing a higher number of fossil inclusions. This has enabled the detailed reconstruction of the ecosystem of a long-vanished tropical forest. Resin from the extinct species *Hymenaea protera* is the source of Dominican amber and probably of most amber found in the tropics. It is not "succinite" but "retinite".

Class II

These ambers are formed from resins with a sesquiterpenoid base, such as cadinene.

Class III

These ambers are polystyrenes.

Class IV

This class is something of a wastebasket; its ambers are not polymerized, but mainly consist of cedrane-based sesquiterpenoids.

Class V

Class V resins are considered to be produced by a pine or pine relative. They comprise a mixture of diterpinoid resins and *n*-alkyl compounds. Their type mineral is *highgate copalite*.

Classification of Baltic Amber (Succinite) Gemstones by the International Amber Association

- *Natural Baltic amber* – gemstone which has undergone mechanical treatment only (for instance: grinding, cutting, turning or polishing) without any change to its natural properties
- *Modified Baltic amber* – gemstone subjected only to thermal or high-pressure treatment, which changed its physical properties, including the degree of transparency and colour, or shaped under similar conditions out of one nugget, previously cut to the required size.
- *Reconstructed (pressed) Baltic amber* – gemstone made of Baltic amber pieces pressed in high temperature and under high pressure without additional components.

- *Bonded Baltic amber* – gemstone consisting of two or more parts of natural, modified or reconstructed Baltic amber bonded together with the use of the smallest possible amount of a colourless binding agent necessary to join the pieces.

Geological Record

The oldest amber recovered dates to the Upper Carboniferous period (320 million years ago). Its chemical composition makes it difficult to match the amber to its producers – it is most similar to the resins produced by flowering plants.

Amber becomes abundant long afterwards, in the Early Cretaceous, 150 million years ago, when it is found in association with insects. The oldest amber with arthropod inclusions comes from the Middle East from Lebanon and Jordan.

This amber is roughly 125–135 million years old and is considered of high scientific value. In Lebanon more than 450 outcrops of Lower Cretaceous amber have been recovered recently among which about 20 outcrops have lead to intensive biological inclusions comprising the oldest respresentatives of several recent families of terrestrial arthropods. Jurassic amber have been found recently in Lebanon as well. Many remarkable insects and spiders were recently discovered in the amber of Jordan (Kaddumi, 2007) including the oldest zorapterans, clerid beetles, umenocoleid roaches, and achiliid planthoppers.

Baltic amber or succinite (historically documented as Prussian amber) is found as irregular nodules in marine glauconitic sand, known as *blue earth,* occurring in the Lower Oligocene strata of Samland in Prussia (Latin: Sambia), in historical sources also referred to as *Glaesaria*. After 1945 this territory around Königsberg was turned into Kaliningrad Oblast, Russia, where it is now systematically mined.

It appears, however, to have been partly derived from yet earlier Tertiary deposits (Eocene); and it occurs also as a derivative phase in later formations, such as glacial drift. Relics of an abundant flora occur as inclusions trapped within the amber while the resin was yet fresh, suggesting relations with the flora of Eastern Asia and the southern part of North America. Heinrich Göppert named the common amber-yielding pine of the Baltic forests *Pinites succiniter*, but as the wood does not seem to differ from that of the existing genus it has been also called *Pinus succinifera*. It is improbable, however, that the production of amber was limited to a single species; and indeed a large number of conifers belonging to different genera are represented in the amber-flora.

Paleontological Significance

Amber is a unique preservational mode, preserving otherwise unfossilizable parts of organisms; as such it is helpful in the reconstruction of ecosystems and organisms.

The chemical composition of the resin is, unfortunately, of limited utility in reconstructing the phylogenetic affinity of the resin producer.

Amber sometimes contains animals or plant matter that became caught in the resin as it was secreted. Insects, spiders and their webs, annelids, frogs, crustaceans, bacteria and amoebae, marine microfossils, wood, flowers and fruit, hair, feathers and other small organisms have been recovered in ambers dating to 130 million years ago.

Use

Amber has been used since antiquity in the manufacture of jewellery and ornaments, and also in folk medicine. Amber also forms the flavoring for akvavit liquor. Amber has been used as an ingredient in perfumes.

Jewellery

Amber has been used since the stone age, from 13,000 years ago. Amber ornaments have been found in Mycenaean tombs and elsewhere across Europe. To this day it is used in the manufacture of smoking and glassblowing mouthpieces. Amber's place in culture and tradition lends it a tourism value; Palanga Amber Museum is dedicated to the mineral.

Amber Frog Violin Bow

The Amber Frog bow by Keith Peck was made in 1996/97. Baltic amber has been used to create the "frog" part of a violin bow. It was commissioned by Gennady Filimonov and made by the late American Master Bowmaker Keith Peck. The Amber Frog/Picture bow (copy of F.N. Voirin), is the first documented amber frog bow (made in 1996–97) that was (and is) a complete success. It is still being played by Gennady Filimonov.

Historic Medicinal Uses

Amber has long been used in folk medicine for its purported healing properties. Amber and extracts were used from the time of

Hippocrates in ancient Greece for a wide variety of treatments through the Middle Ages and up until the early twentieth century.

Scent of Amber and Amber Perfumery

In ancient China it was customary to burn amber during large festivities. If amber is heated under the right conditions, oil of amber is produced, and in past times this was combined carefully with nitric acid to create "artificial musk" – a resin with a peculiar musky odor. Although when burned, amber does give off a characteristic "pinewood" fragrance, modern products, such as perfume, do not normally use actual amber. This is due to the fact that fossilized amber produces very little scent. In perfumery, scents referred to as "amber" are often created and patented to emulate the opulent golden warmth of the fossil. The modern name for amber is thought to come from the Arabic word, ambar, meaning ambergris. Ambergris is the waxy aromatic substance created in the intestines of sperm whales and was used in making perfumes both in ancient times as well as modern. The scent of amber was originally derived from emulating the scent of ambergris and/or labdanum but due to the endangered status of the sperm whale the scent of amber is now largely derived from labdanum. The term "amber" is loosely used to describe a scent that is warm, musky, rich and honey-like, and also somewhat oriental and earthy. It can be synthetically created or derived from natural resins. When derived from natural resins it is most often created out of labdanum. Benzoin is usually part of the recipe. Vanilla and cloves are sometimes used to enhance the aroma.

"Amber" perfumes may be created using combinations of labdanum, benzoin resin, copal (itself a type of tree resin used in incense manufacture), vanilla, Dammara resin and/or synthetic materials.

Alum

Alum is both a specific chemical compound and a class of chemical compounds. The specific compound is the hydrated potassium aluminium sulfate (potassium alum) with the formula $KAl(SO_4)_2 \cdot 12H_2O$. The wider class of compounds known as alums have the related empirical formula, $AB(SO_4)_2 \cdot 12H_2O$.

Chemical Properties

Alums are useful for a range of industrial processes. They are soluble in water; have an astringent, acid, and sweetish taste; react acid to litmus; and crystallize in regular octahedra. When heated they liquefy; and if the heating is continued, the water of crystallization is driven off, the salt froths and swells, and at last an amorphous powder remains.

Figure 5 : Bulk alum

Industrial Uses

Potassium alum is the common alum of commerce, although soda alum, ferric alum, and ammonium alum are manufactured.

Alum is also used in purification of drinking water in industries. In a holding tank, some alum (phitkari) is added to the water so that the negatively charged light colloidal parts stick together and get heavy (flocculate) when alum makes the colloidal particles neutralized by making its aluminum ions get loaded with the colloidal parts. When the colloidal parts get heavy they can be easily separated from the tank prior to further filtration and disinfection of the water.

Cosmetic

- Alum in block form (usually potassium alum) is used as a blood coagulant.
- Styptic pencils containing aluminium sulfate or potassium aluminium sulfate are used as astringents to prevent bleeding from small shaving cuts.
- Alum was used as a base in skin whiteners and treatments during the late 16th century. A recipe for one such compound

was given thus: "For the Freckles which one getteth by the heat of the Sun: Take a little Allom beaten small, temper amonst it a well brayed white of an egg, put it on a milde fire, stirring it always about that it wax not hard, and when it casteth up the scum, then it is enough, wherewith anoint the Freckles the space of three dayes: if you will defend your self that you get no Freckles on the face, then anoint your face with the whites of eggs." —Christopher Wirzung, *General Practise of Physicke,* 1654.

- Alum may be used in depilatory waxes used for the removal of body hair, or applied to freshly waxed skin as a soothing agent.
- In the 1950s, men sporting crewcut or flattop hairstyles sometimes applied alum to their front short hairs as an alternative to pomade. When the hair dried, it would stay up all day.
- Alum's antibacterial properties contribute to its traditional use as an underarm deodorant. It has been used for this purpose in Europe; Mexico; Thailand, where it is called Sarn-Som; throughout Asia; and in the Philippines, where it is called Tawas. Today, potassium alum is sold commercially for this purpose as a "deodorant crystal," often in a protective plastic case.

Medicinal

- Alum is used in many subunit vaccines as an adjuvant to enhance the body's response to immunogens. Such vaccines include hepatitis A, hepatitis B, and DTaP.
- Alum in powder or crystal form, or in styptic pencils, is sometimes applied to cuts to prevent or treat infection.
- Powdered alum is commonly cited as a home remedy for canker sores.
- Preparations containing alum are used by pet owners to stem bleeding associated with animal injuries caused by improper nail clipping.
- Alum is listed as an ingredient of some brands of toothpaste or toothpowder.
- Alum powder dissolved in 5 parts water has been used to shrink hemorrhoids and stop them from bleeding.

Culinary

- Alum powder, found in the spice section of many grocery stores, may be used in pickling recipes as a preservative to maintain fruit and vegetable crispness.
- Alum is used as the acidic component of some commercial baking powders.
- Alum was used by bakers in England during the 1800s to make bread whiter. In 1875, the *Sale of Food and Drugs Act* prevented this and other adulterations.

Flame Retardant

- Solutions containing alum may be used to treat cloth, wood, and paper materials to increase their resistance to fire.
- Alum is also a component of foamite.
- Alum is also used in fire extinguishers to smother chemical and oil fires.

Chemical Flocculant

- Alum is used to clarify water by neutralizing the electrical double layer surrounding very fine suspended particles, allowing them to flocculate (stick together). After flocculation, the particles will be large enough to settle and can be removed.
- Alum may be used to increase the viscosity of a ceramic glaze suspension; this makes the glaze more readily adherent and slows its rate of sedimentation.
- Alum is an ingredient in some recipes for homemade modelling compounds intended for use by children. (These are often called "play clay" or "play dough" for their similarity to "Play-Doh", a trademarked product marketed by American toy manufacturer Hasbro).

History

In Antiquity

The word "alumen" occurs in Pliny's *Natural History*. In the 52nd chapter of his 35th book, he gives a detailed description. By comparing this with the account of stupteria given by Dioscorides in the 123rd chapter of his 5th book, it is obvious that the two are identical. Pliny informs us that alumen was found naturally in the earth. He calls it *salsugoterrae*. Different substances were distinguished by the name of "alumen"; but they were all characterized by a certain degree of astringency, and were all employed in dyeing and medicine, the light-coloured alumen being useful in brilliant dyes, the dark-coloured only in dyeing black or very dark colours. One species was a liquid, which was apt to be adulterated; but when pure it had the property of blackening when added to pomegranate juice. This property seems to characterize a solution of iron sulfate in water; a solution of ordinary (potassium) alum would possess no such property. Pliny says that there is another kind of alum that the Greeks call *schistos*. It forms in white threads upon the surface of certain stones. From the name schistos, and the mode of formation, it appears that this species was the salt which forms spontaneously on certain salty minerals, as alum slate and bituminous shale, and which consists chiefly of sulfates of iron and aluminium. Possibly in certain places the iron sulfate may have been nearly wanting, and then the salt would be white, and would answer, as Pliny says it did, for dyeing bright colours. Several other species of alumen are described by Pliny, but we are unable to make out to what minerals he alludes.

The alumen of the ancients, then, was not always the same as the alum of the moderns. They certainly knew how to produce alum from alunite, as this process is archaeologically attested on the island Lesbos. This site was abandoned in the 7th century but dates back at least to the 2nd century AD. Native alumen from Melos appears to have been a mixture mainly of alunogen ($Al_2(SO_4)_3.17H_2O$) with alum and other minor sulfates. The western desert of Egypt was a major source of alum substitutes in antiquity. These evaporites were mainly $FeAl_2(SO_4)_4.22H_2O$, $MgAl_2(SO_4)_4.22H_2O$, $NaAl(SO_4)_2.6H_2O$, $MgSO_4.7H_2O$ and $Al_2(SO_4)_3.17H_2O$. Any contamination with iron sulfate was greatly disliked as this darkened and dulled dye colours. They were acquainted with a variety of substances of varying degrees of purity by the names of misy, sory, and chalcanthum. As alum and green vitriol were applied to a variety of substances in common, and as both are distinguished by a sweetish and astringent taste, writers, even after the discovery of alum, do not seem to have discriminated the two salts accurately from each other. In the writings of the alchemists we find the words misy, sory, chalcanthum applied to alum as well as to iron sulfate; and the name atramentum sutorium, which ought to belong, one would suppose, exclusively to green vitriol, applied indifferently to both. Various minerals are employed in the manufacture of alum, the most important being alunite, alum schist, bauxite and cryolite.

Alchemical and Later Discoveries and Uses

In the 18th century, J. H. Pott and Andreas Sigismund Marggraf demonstrated that alumina was a constituent. Pott in his *Lithogeognosia* showed that the precipitate obtained when an alkali is poured into a solution of alum is quite different from lime and chalk, with which it had been confounded by G.E. Stahl. Marggraf showed that alumina is one of the constituents of alum, but that this earth possesses peculiar properties, and is one of the ingredients in common clay. He also showed that crystals of alum can be obtained by dissolving alumina in sulfuric acid and evaporating the solutions, and when a solution of potash or ammonia is dropped into this liquid, it immediately deposits perfect crystals of alum.

Torbern Bergman also observed that the addition of potash or ammonia made the solution of alumina in sulfuric acid crystallize, but that the same effect was not produced by the addition of soda or of lime, and that potassium sulfate is frequently found in alum.

After M.H. Klaproth had discovered the presence of potassium in leucite and lepidolite, it occurred to L.N. Vauquelin that it was probably an ingredient likewise in many other minerals. Knowing that alum cannot be obtained in crystals without the addition of potash, he began to suspect that this alkali constituted an essential ingredient in the salt, and in 1797 he published a dissertation demonstrating that alum is a double salt, composed of sulfuric acid, alumina, and potash. Soon after, J.A. Chaptal published the analysis of four different kinds of alum, namely, Roman alum, Levant alum, British alum and alum manufactured by himself. This analysis led to the same result as Vauquelin.

Early Uses in Industry

Egyptians reportedly used the coagulant alum as early as 1500 BC to reduce the visible cloudiness (turbidity) in the water. Alum was imported into England mainly from the Middle East, and, from the late 15th century onwards, the Papal States for hundreds of years. Its use there was as a dye-fixer (mordant) for wool (which was one of England's primary industries, the value of which increased significantly if dyed). These sources were unreliable, however, and there was a push to develop a source in England especially as imports from the Papal States were ceased following the excommunication of King Henry VIII. With state financing, attempts were made throughout the 16th century, but without success until early on in the 17th century. An industry was founded in Yorkshire to process the shale which contained the key ingredient, aluminium sulfate, and made an important contribution to the Industrial Revolution.

One of the oldest historic sites for the production of alum from shale and human urine are the Peak alum works in Ravenscar, North Yorkshire.

Alum was also used for water treatment by Indians for thousands of years. Ayurveda describes sphatika as an astringent, haemostatic, antiseptic. It has anti-inflammatory, anti-pyretic and antibiotic properties. Sphatika's use in treating tonsillitis has been referred in ancient Ayurvedic texts. Sphatika is used internally as well as externally.

Production

From Alunite

In order to obtain alum from alunite, it is calcined and then exposed to the action of air for a considerable time. During this exposure it is kept continually moistened with water, so that it ultimately falls to a very fine powder. This powder is then lixiviated with hot water and sulfuric acid, the liquor decanted, and the alum allowed to crystallize.

The alum schists employed in the manufacture of alum are mixtures of iron pyrite, aluminium silicate and various bituminous substances, and are found in upper Bavaria, Bohemia, Belgium, and Scotland. These are either roasted or exposed to the weathering action of the air.

In the roasting process, sulfuric acid is formed and acts on the clay to form aluminium sulfate, a similar condition of affairs being produced during weathering. The mass is now systematically extracted with water, and a solution of aluminium sulfate of specific gravity 1.16 is prepared. This solution is allowed to stand for some time (in order that any calcium sulfate and basic ferric sulfate may separate), and is then evaporated until ferrous sulfate crystallizes on cooling; it is then drawn off and evaporated until it attains a specific gravity of 1.40. It is now allowed to stand for some time, decanted from any sediment, and finally mixed with the calculated quantity of potassium sulfate, well agitated, and the alum is thrown down as a finely divided precipitate of alum meal. If much iron should be present in the shale then it is preferable to use potassium chloride in place of potassium sulfate.

From Clays or Bauxite

In the preparation of alum from clays or from bauxite, the material is gently calcined, then mixed with sulfuric acid and heated gradually to boiling; it is allowed to stand for some time, the clear solution drawn off and mixed with acid potassium sulfate and allowed to crystallize.

When cryolite is used for the preparation of alum, it is mixed with calcium carbonate and heated. By this means, sodium aluminate is formed; it is then extracted with water and precipitated either by sodium bicarbonate or by passing a current of carbon dioxide through the solution. The precipitate is then dissolved in sulfuric acid, the requisite amount of potassium sulfate added and the solution allowed to crystallize.

Types

Figure 6: Alum crystal with small amount of chrome alum to give a slight violet color

Potash Alum

Aluminum potassium sulfate, potash alum, $KAl(SO_4)_2 \cdot 12H_2O$ is used as an astringent and antisepsis in various food preparation processes such as pickling and fermentation and as a flocculant for water purification among other things.

A common method of producing potash alum is leaching of alumina from bauxite which is then reacted with potassium sulfate. As a naturally occurring mineral potash alum is known as kalunite. Other potassium aluminium sulfate minerals are alunite ($KAl(SO_4)_2 \cdot 2Al(OH)_3$) and kalinite ($KAl(SO_4)_2 \cdot 11H_2O$).

The molecular weight of potash alum is 474 g/mole.

Soda Alum

Soda alum, $NaAl(SO_4)_2 \cdot 12H_2O$, mainly occurs in nature as the mineral mendozite. It is very soluble in water, and is extremely difficult to purify. In the preparation of this salt, it is preferable to mix the component solutions in the cold, and to evaporate them at a temperature not exceeding 60 °C. 100 parts of water dissolve 110 parts of sodium alum at 0 °C, and 51 parts at 16 °C. Soda alum is used in the acidulent of food as well as in the manufacture of baking powder.

Ammonium Alum

Ammonium alum, $NH_4Al(SO_4)_2 \cdot 12H_2O$, a white crystalline double sulfate of aluminium, is used in water purification, in vegetable glues, in porcelain cements, in deodorants (though potassium alum is more commonly used), in tanning, dyeing and in fireproofing textiles.

Chrome Alum

Chrome alum, $KCr(SO_4)_2 \cdot 12H_2O$, a dark violet crystalline double sulfate of chromium and potassium, was used in tanning.

Selenate-containing Alums

Alums are also known that contain selenium in place of sulfur in the sulfate anion, making selenate (SeO_4^{2-}) instead. They are called selenium-or selenate-alums. They are strong oxidizing agents.

Aluminium Sulfate

Aluminium sulfate is referred to as papermaker's alum. Although reference to this compound as alum is quite common in industrial communication, it is not regarded as technically correct. Its properties are quite different from those of the set of alums. Most industrial flocculation done with alum is actually aluminum sulfate.

Crystal Chemistry

Double sulfates with the general formula $A_2SO_4 \cdot B_2(SO_4)_3 \cdot 24H_2O$, are known where A is a monovalent cation such as sodium, potassium, rubidium, caesium, or thallium(I), or a compound cation such as ammonium (NH_4^+), methylammonium ($CH_3NH_3^+$), hydroxylammonium

($HONH_3^+$) or hydrazinium ($N_2H_5^+$), *B* is a trivalent metal ion, such as aluminium, chromium, titanium, manganese, vanadium, iron (III), cobalt(III), gallium, molybdenum, indium, ruthenium, rhodium, or iridium.

The specific combinations of univalent cation, trivalent cation, and anion depends on the sizes of the ions. For example, unlike the other alkali metals the smallest one, lithium, does not form alums, and there is only one known sodium alum. In some cases, solid solutions of alums occur.

Alums crystallize in one of three different crystal structures. These classes are called α-, β-and γ-alums.

Solubility

The solubility of the various alums in water varies greatly, sodium alum being readily soluble in water, while caesium and rubidium alums are only sparingly soluble. The various solubilities are shown in the following table.

At temperature *T*, 100 parts water dissolve:

T	*Ammonium Alum*	*Potassium Alum*	*Rubidium Alum*	*Caesium Alum*
0 °C	2.62	3.90	0.71	0.19
10 °C	4.50	9.52	1.09	0.29
50 °C	15.9	44.11	4.98	1.235
80 °C	35.20	134.47	21.60	5.29
100 °C	70.83	357.48		

Related Compounds

In addition to the alums, which are dodecahydrates, double sulfates and selenates of univalent and trivalent cations occur with other degrees of hydration. These materials may also be referred to as alums, including the undecahydrates such as mendozite and kalinite, hexahydrates such as guanidinium ($CH_6N_3^+$) and dimethylammonium ($(CH_3)_2NH_2^+$) "alums", tetrahydrates such as goldichite, monohydrates such as thallium plutonium sulfate and anhydrous alums (yavapaiites). These classes include differing, but overlapping, combinations of ions.

A pseudo alum is a double sulfate of the typical formula $ASO_4{\cdot}B_2(SO_4)_3{\cdot}22H_2O$, where A is a divalent metal ion, such as cobalt (wupatkiite), manganese (apjohnite), magnesium (pickingerite) or iron (halotrichite or feather alum), and B is a trivalent metal ion.

A Tutton salt is a double sulfate of the typical formula $A_2SO_4 \cdot BSO_4 \cdot 6H_2O$, where A is a univalent cation, and B a divalent metal ion.

Double sulfates of the composition $A_2SO_4 \cdot 2BSO_4$, where A is a univalent cation and B is a divalent metal ion are referred to as langbeinites, after the prototypical potassium magnesium sulfate.

In popular culture

Much use was made of the supposed properties of alum as a comedy gag in films, primarily in the 1920s and 1930s. In a typical situation it would be introduced by accident or intent into foodstuffs, with ingestion causing the victim's mouth to assume a tight pucker. Speech was usually difficult or impossible. In animation, cartoon physics could magnify the effect-the victim's head might shrink (as in the 1949 Bugs Bunny cartoon *Long-Haired Hare*) or the voice alter to a shrill squeak.

Chapter 4

Important Minerals

Amazonite

Amazonite (sometimes called "Amazon stone") is a green variety of microcline feldspar.

The name is taken from that of the Amazon River, from which certain green stones were formerly obtained, but it is doubtful whether green feldspar occurs in the Amazon area.

Amazonite is a mineral of limited occurrence. Formerly it was obtained almost exclusively from the area of Miass in the Ilmen mountains, 50 miles southwest of Chelyabinsk, Russia, where it occurs in granitic rocks. More recently, high-quality crystals have been obtained from Pike's Peak, Colorado, where it is found associated with smoky quartz, orthoclase, and albite in a coarse granite or pegmatite. Crystals of amazonite can also be found in Crystal Park, El Paso County, Colorado. Some other localities in the United States yield amazonite, and it is also found in pegmatite in Madagascar and in Brazil.

Because of its bright green colour when polished, amazonite is sometimes cut and used as a gemstone, although it is easily fractured.

For many years, the source of amazonite's colour was a mystery. Naturally, many people assumed the colour was due to copper because copper compounds often have blue and green colours. More recent studies suggest that the blue-green colour results from small quantities of lead and water in the feldspar.

Amethyst

Amethyst is a violet variety of quartz often used in jewellery. The name comes from the Ancient Greek *a*-("not") and *ì èõóôïò methustos* ("intoxicated"), a reference to the belief that the stone protected its owner from drunkenness; the ancient Greeks and Romans wore amethyst and made drinking vessels of it in the belief that it would prevent intoxication. It is one of several forms of quartz. Amethyst is the traditional birthstone for February.

Structure

Amethyst is the purple variety of quartz (SiO_2), containing an impurity of iron, which gives the violet colour to the mineral. The hardness of the mineral is the same as quartz's, thus it is suitable for use in jewellery.

Hue and Tone

Amethyst occurs in primary hues from a light pinkish violet to a deep purple. Amethyst may exhibit one or both secondary hues, red and blue. The ideal grade is called "Deep Siberian" and has a primary purple hue of around 75–80%, 15–20% blue and (depending on the light source) red secondary hues. Green quartz is sometimes incorrectly called green amethyst, which is an actual misnomer and not an acceptable name for the material, the proper terminology being prasiolite. It is actually against FTC Guidelines to call prasiolite *green amethyst*. Other names for green quartz are vermarine, greened amethyst, or lime citrine.

Of very variable intensity, the colour of amethyst is often laid out in stripes parallel to the final faces of the crystal. One aspect in the art of lapidary involves the correct cutting to place the colour correctly in order to make the tone of the cut stones homogeneous. The fact that sometimes only a thin surface layer of violet colour is present in the stone, or that the colour is not homogeneous often makes for a difficult cutting.

The colour of amethyst has been demonstrated to result from substitution of trivalent iron (Fe^{3+}) in the structure. When heated, amethyst may turm yellow-orange, yellow-brown, or dark brownish and may resemble citrine.

History

Amethyst was used as a gemstone by the ancient Egyptians and was largely employed in antiquity for intaglio engraved gems.

The Greeks believed amethyst gems could prevent intoxication, while medieval European soldiers wore amethyst amulets as protection in battle – the reason for this being that amethysts are believed to heal people and keep them cool-headed. Beads of amethyst were found in Anglo-Saxon graves in England. A large geode, or "amethyst-grotto", from near Santa Cruz in southern Brazil was presented at the 1902 exhibition in Düsseldorf, Germany.

In the 19th century, the colour of amethyst was attributed to the presence of manganese. However, since it is capable of being greatly altered and even discharged by heat, the colour was believed by some authorities to be from an organic source. Ferric thiocyanate has been suggested, and sulfur was said to have been detected in the mineral.

Synthetic Amethyst

Synthetic amethyst is thus man-made by gamma-ray, x-ray or electron beam irradiation of clear quartz which has been first doped with ferric impurities. On exposure to heat, the irradiation effects can be partially cancelled and amethyst generally becomes yellow or even green, and much of the citrine, cairngorm, or yellow quartz of jewellery is said to be merely "burnt amethyst".

Synthetic amethyst is made to imitate the best quality amethyst. Its chemical and physical properties are so similar to that of natural amethyst that it can not be differentiated with absolute certainty without advanced gemnological testing (which is often cost-prohibitive). There is one test based on "Brazil law twinning" (a form of quartz twinning where right and left hand quartz structures are combined in a single crystal) which can be used to identify synthetic amethyst rather easily. It is possible to synthesize twinned amethyst, but this type is not available in large quantities in the market.

Mythology

The Greek word "amethystos" may be translated as "not drunken", from Greek *a-*, "not" + *methustos*, "intoxicated". Amethyst was considered to be a strong antidote against drunkenness, which is why wine goblets were often carved from it. In Greek mythology, Dionysus, the god of intoxication, and of wine, was pursuing a maiden named Amethystos, who refused his affections. Amethystos prayed to the gods to remain chaste, a prayer which the goddess Artemis answered, transforming her into a white stone. Humbled by Amethystos's desire to remain chaste, Dionysus poured wine over the stone as an offering, dyeing the crystals purple.

Variations of the story include that Dionysus had been insulted by a mortal and swore to slay the next mortal who crossed his path, creating fierce tigers to carry out his wrath. The mortal turned out to be a beautiful young woman, Amethystos, who was on her way to pay tribute to Artemis. Her life was spared by Artemis, who transformed the maiden into a statue of pure crystalline quartz to protect her from the brutal claws. Dionysus wept tears of wine in remorse for his action at the sight of the beautiful statue. The god's tears then stained the quartz purple. Another variation involves the titan Rhea presenting Dionysus with the amethyst stone to preserve the wine-drinker's sanity.

Geographic Distribution

Amethyst is produced in abundance from the state of Minas Gerais in Brazil where it occurs in large geodes within volcanic rocks. Many of the hollow agates of southwestern Brazil and Uruguay contain a crop of amethyst crystals in the interior. Artigas, Uruguay and neighboring Brazilian state Rio Grande do Sul are large world producers exceeding in quantity Minas Gerais, as well as Mato Grosso, Espirito Santo, Bahia, and Ceará states, all amethyst producers of importance in Brazil.

Figure 1 : An amethyst geode that formed when large crystals grew in open spaces inside the rock.

It is also found and mined in South Korea. The largest opencast amethyst vein in the world is in Maissau, Lower Austria. Much fine amethyst comes from Russia, especially from near Mursinka in the Ekaterinburg district, where it occurs in drusy cavities in granitic rocks. Many localities in south India yield amethyst. One of the largest global amethyst producers is Zambia in southern Africa with an annual production of about 1000 tonnes.

Amethyst occurs at many localities in the United States. Among these may be mentioned: the Mazatzal Mountain region in Gila and Maricopa Counties, Arizona; Red Feather Lakes, near Ft Collins, Colorado; Amethyst Mountain, Texas; Yellowstone National Park; Delaware County, Pennsylvania; Haywood County, North Carolina; Deer Hill and Stow, Maine and in the Lake Superior region of Minnesota, Wisconsin, Michigan, and Ontario in Canada.

Amethyst is relatively common in Ontario, and in various locations throughout Nova Scotia. The largest amethyst mine in North America is located in Thunder Bay, Ontario.

Value

Up until the 18th century, amethyst was included in the cardinal, or most valuable, gemstones (along with diamond, sapphire, ruby, and emerald). However, since the discovery of extensive deposits in locations such as Brazil, it has lost most of its value.

Collectors look for depth of colour, possibly with red flashes if cut conventionally. The highest grade amethyst (called "Deep Russian") is exceptionally rare and therefore, when one is found, its value is dependent on the demand of collectors. It is, however, still orders of magnitude lower than the highest grade sapphires or rubies (padparadscha sapphire or "pigeon's blood" ruby).

Ammolite

Ammolite is a rare and valuable opal-like organic gemstone found primarily along the eastern slopes of the Rocky Mountains of North America.

It is made of the fossilized shells of ammonites, which in turn are composed primarily of aragonite, the same mineral that makes up nacreous pearls. It is one of several biogenic gemstones; others include amber and pearl. In 1981, ammolite was given official gemstone status by the World Jewellery Confederation, the same year commercial mining of ammolite began. It was designated the official gemstone of the Province of Alberta in 2004 and the official gemstone of the City of Lethbridge in 2007.

Ammolite is also known as *aapoak* (Kainah for "small, crawling stone"), gem ammonite, calcentine, and korite. The latter is a trade name given to the gemstone by the Alberta-based mining company Korite International, the first and largest commercial producer of ammolite.

Properties

The chemical composition of ammolite is variable, and aside from aragonite may include calcite, silica, pyrite, or other minerals. The shell itself may contain a number of trace elements, including: aluminium; barium; chromium; copper; iron; magnesium; manganese; strontium; titanium; and vanadium. Its crystallography is orthorhombic. Its hardness is 4.5–5.5, quite soft for a gemstone, and its specific gravity is 2.60–2.85. The refractive index of Canadian material (as measured via sodium light, 589.3 nm) is as follows: α 1.522; β 1.672–1.673; γ 1.676–1.679; biaxial negative. Under ultraviolet light, ammolite may fluoresce a mustard yellow.

An iridescent opal-like play of colour is shown in fine specimens, mostly in shades of green and red; all the spectral colours are possible, however. The iridescence is due to the microstructure of the aragonite: unlike most other gems, whose colours come from light absorption, the iridescent colour of ammolite comes from interference with the light that rebounds from stacked layers of thin platelets that make up the aragonite.

The thicker the layers, the more reds and greens are produced; the thinner the layers, the more blues and violets predominate. Reds and greens are the most commonly seen colours, owing to the greater fragility of the finer layers responsible for the blues. When freshly quarried, these colours are not especially dramatic; the material requires polishing and possibly other treatments in order to reveal the colours' full potential.

The ammolite itself is actually a very thin sheet, ca. 0.5–0.8 millimeters (0.02–0.03 inches) in thickness. Rarely is ammolite without its matrix, which is typically a grey to brown shale, chalky clay, or limestone. So-called "frost shattering" is common; exposed to the elements and compressed by sediments, the thin ammolite tends to crack and flake; prolonged exposure to sunlight can also lead to bleaching. The cracking results in a tessellated appearance, sometimes described as a "dragon skin" or "stained glass window" pattern. Ammolite mined from deeper deposits may be entirely smooth or with a rippled surface.

Occasionally a complete ammonite shell is recovered with its structure well-preserved: fine, convoluted lines delineate the shell chambers, and the overall shape is suggestive of a nautilus. While these shells may be as large as 90 centimeters (35.5 inches) in diameter, the iridescent ammonites (as opposed to the pyritized variety) are typically much smaller. Most fossilized shells have had their aragonite pseudomorphously replaced by calcite or pyrite, making the presence of ammolite particularly uncommon.

Formation, Occurrence, and Extraction

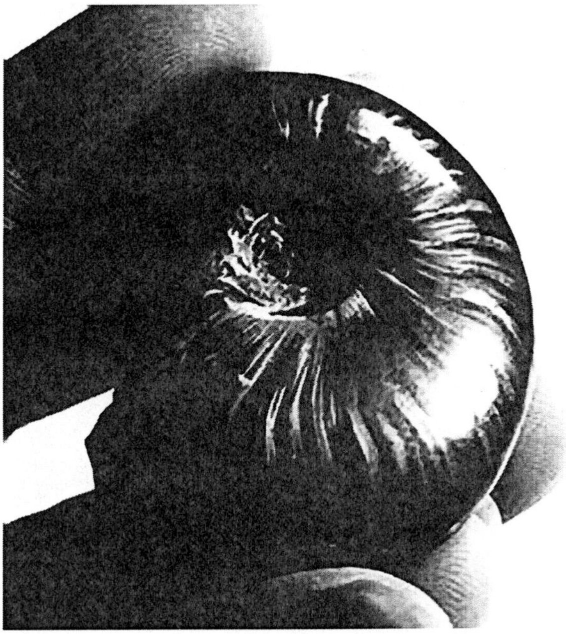

Figure 2 : An iridescent ammonite from Madagascar

Ammolite comes from the fossil shells of the Upper Cretaceous disk-shaped ammonites *Placenticeras meeki* and *Placenticeras intercalare*, and (to a lesser degree) the cylindrical baculite, *Baculites compressus*. Ammonites were cephalopods, or squid-like creatures, that thrived in tropical seas until becoming extinct along with the dinosaurs at the end of the Mesozoic era.

The ammonites that form ammolite inhabited a prehistoric, inland subtropical sea that bordered the Rocky Mountains—this area is known today as the Cretaceous or Western Interior Seaway. As the seas receded, the ammonites were buried by layers of bentonite sediment. This sediment preserved the aragonite of their shelled remains, preventing it from converting to calcite.

Significant deposits of gem-quality ammolite are only found in the Bearpaw formation that extends from Alberta to Saskatchewan in Canada and south to Montana in the USA. The best grade of gem quality ammolite is along high energy river systems on the eastern slopes of the Rockies in southern Alberta. Most commercial mining operations have been conducted along the banks of the St. Mary River, in an area south of and between the town of Magrath and the city of Lethbridge.

Roughly half of all ammolite deposits are contained within the Kainah (Kainaiwa) reserve, and its inhabitants play a major role in ammolite mining. Since its founding in 1979, Korite International has operated primarily within the reservation. The company maintains an agreement with the Kainah (Blood) tribe, with Korite International paying the tribe royalties based on how much land the company has mined.

Figure 3 : Another view of Korite International's open-pit mining operations in Alberta, Canada.

Commercial extraction is mechanized but fairly basic: shallow open pits are dug with a backhoe and the excavated material is screened for its potential gem contents. The pits are further examined by hand, and commercial production is supplemented by individuals who sell their surface-picked findings to Korite International and several other producers. Approximately 50% of the ammolite mined is suitable for jewellery.

The ammolite deposits are stratified into several layers: the shallowest of these layers, named the "K Zone", lies some 15 meters below the surface and extends 30 meters down. The ammolite within this layer is covered by siderite concretions and is usually cracked—this is the *crush* material. It is the most common and (generally speaking) the least valuable of ammolite. Beginning twenty meters below the crush material is the "Blue Zone"; ammolite from this zone, which extends 65 meters, is usually compressed with a thin layer of pyrite rather than siderite concretions. This is the *sheet* material; due to its depth it is rarely mined. It is also much less fractured, and therefore the more valuable type of ammolite.

As of 2003, Korite International had only mined 30 acres (120,000 m^2) of the Kainah deposit. As part of their agreement with the tribe, the company must refill areas once exhausted to mitigate environmental degradation. Korite International's operations net the Kainah tribe approximately CAD 150,000–200,000 per year.

The company employs over 60 people (many of them Kainah), and accounts for approximately 90 percent of world gem ammolite production. Prospectors who wish to mine ammolite deposits on Crown land must apply to the Alberta Department of Energy for a lease. These leases are not regularly offered; as of 2004, there was a CAD 625 application fee, with an annual rental fee of CAD $3.50 per hectare.

Gemstone Quality

The quality of gem ammolite is communicated via a letter grade system, from best to least best: AA; A+; A; and A-. However, this system is not yet standardized and some vendors may use their own systems. The grade and therefore the value of an ammolite gemstone is determined by the following criteria:

The number of primary colours : A large array of colour is displayed in ammolite, including all the spectral colours found in nature. Red and green are far more common than blue or purple due to the latter's fragility. There are also certain hues, like crimson or violet or gold, which are derived from a combination of the primary colours, that are the rarest and in highest demand. The most valuable grades have roughly equal portions of three or more primary colours or 1–2 bright and even colours, with the lowest grades having one comparatively dull colour predominant.

The way the colours play (chromatic shift and rotational range) : *Chromatic shift* is how the colours vary with the angle of viewing and the angle of light striking the gemstone; in higher grades this variation is almost prismatic in its scope, while lower grades show very little variation. *Rotational range* is how far the specimen can be turned while maintaining its play of colour; the best rotate 360 degrees uncompromised, while lesser stones may exhibit highly directional colours that are only visible within a narrow rotational range, down to 90° or less. Intermediate grades have ranges of 240–180°.

Brightness of colours (iridescence) : The brightness of colours and their iridescence is essentially dependent on how well-preserved the nacreous shell is, and how fine and orderly the layers of aragonite are. The quality of the polish is also a factor. The "dragon skin" cracking usually hinders its value; the most prized ammolite is the *sheet* type and has broad, uninterrupted swathes of colour similar to the "broad flash" category of opal. The matrix is not visible in finer grades, and there should be no foreign minerals breaking up or diminishing the iridescence.

The thickness of the ammolite layer is also an important factor: after polishing, the ammolite is only 0.1–0.3 millimeters thick. The rarest and most valuable are thick enough to stand alone, with only a thin portion of its original matrix (not exceeding 1.5 mm); but the vast majority require some sort of supportive backing. Other treatments are also commonly undertaken; all other factors being equal, the less treatment an ammolite gem has received, the more valuable it is. Calibrated stones—that is, stones fashioned into standard dimensions that will fit most jewellery settings—may also command a higher price.

Ammolite is sold for USD $30 to 65 per carat (150 to 325 $/g).

Treatments

Although fully mineralized and containing no water—and therefore not subject to dehydration and subsequent crazing as seen in opal—ammolite is often damaged due to environmental exposure. The thin, delicate sheets in which ammolite occurs are also problematic; for these reasons, most material is impregnated with a clear epoxy or other synthetic resin to stabilize the flake-prone ammolite prior to cutting. Although the tessellated cracking cannot be repaired, the epoxy prevents further flaking and helps protect the relatively soft surface from scratching. The impregnation process was developed over a number of years by Korite International in partnership with the Alberta Research Council. Impregnated and epoxy-coated ammolite first entered the market in 1989 and the treatment significantly increased the availability of the gem.

Because the ammolite layer is usually mere fractions of a millimeter in thickness, most ammolite gems are in fact *composite* stones: these usually take the form of two-part *doublets*, with the ammolite layer adhered to a dark backing material. This is usually the matrix or mother rock from which the ammolite was quarried; black onyx or glass could also be used as backing. In composites where the ammolite layer is exceptionally thin, a third component is used: this constitutes a *triplet*, with a durable and transparent convex topping piece. This cap may be either synthetic spinel, synthetic corundum, synthetic quartz, or in lower-end productions, glass. The convex cap acts as a lens and has the effect of enhancing the ammolite's iridescent display.

The detection of these treated and composite stones is relatively simple via inspection with a loupe; however, certain jewellery setting styles—such as those with closed backs—can complicate things. A triplet can be identified by inspecting the stone in profile; the top of the stone can then be seen to be domed and transparent, with no play of colour. If the dome is made of glass, bubbles, swirl marks, and scratches may be present; the harder synthetic materials are optically flawless.

Although the vast majority of commercial-grade ammolite has been treated in some way, a small fraction of production requires no treatment other than cutting and polishing. Ideally, any treatments should be disclosed at the time of sale.

Imitations

Ammolite is neither easily nor often imitated; however, a few materials have a passing resemblance that may deceive the unfamiliar. These include: labradorite (also known as spectrolite), an iridescent feldspar also of Canadian origin; and broad-flash black opal. Neither are convincing substitutes, and the latter is actually of greater value than ammolite. Indeed, ammolite is often used as an imitation of black opal. An even less convincing possibility is Slocum stone, a common glass-based imitation of opal. Blues and purples are much more pervasive in labradorite, and in both it and opal the play of colour is seen to roll across the stone unlike the comparatively restricted play of colour in ammolite. In Slocum stone, the play of colour takes the form of tinsel-like patches. The visible structure is also considerably different; in the imitations, the body of the stone is transparent to translucent from certain angles, whereas ammolite is entirely opaque.

Gemologically speaking, ammolite can be grouped with the shell-based marbles. This group includes *lumachella* or "fire marble", a similarly iridescent marble composed of fossilized clam and snail shells. Found in Italy and Austria, lumachella is rarely if ever used in jewellery; rather, it is used as a decorative facing stone or in mosaics. The iridescence of lumachella is fragmentary and not nearly as brilliant as that of ammolite. Despite these differences, lumachella may be considered synonymous with ammolite in some circles.

The predominantly blue-green iridescent shell of abalone (or paua; genus *Haliotis*) is one last possible imitation. Abalone shell is inexpensive and plentiful owing to the commercial mariculture of these gastropods for their meat. The shell's structure is distinctive: sinuous bands of blue, green, and rose iridescence are delineated by dark brown lines of conchiolin, a proteinaceous material that holds the shell together.

The luster of abalone shell is silky rather than the near vitreous luster of polished ammolite, and the colours of the two materials do not closely approximate. However, some abalone shell has been dyed and given a transparent cap of synthetic quartz, forming a doublet in the same fashion as ammolite. These doublets are perhaps the most deceptive, and have also been used to imitate opal. Under magnification most abalone doublets will show dye concentrated along certain areas and air bubbles trapped at the shell-quartz interface.

Use in jewellery

Compared to most other gems, ammolite has a rather scant history of use; it did not begin to garner interest in Western society until the 1970s after entering the market (to a limited degree) in 1969. The Blackfeet tribe know ammolite as *iniskim*, meaning "buffalo stone", and have long believed it to possess amuletic powers; specifically, the gem is believed to aid in the buffalo hunt, and to draw the buffalo within tracking distance. The Blackfeet also believe ammolite to possess healing powers and incorporate the gem into their medicine bundles for use in ceremonies.

In the late 1990s, practitioners of Feng Shui began to promote ammolite as an "influential" stone with what they believe is the power to enhance well-being and detoxify the body by improving its flow of energy or "chi". Named the "Seven Colour Prosperity Stone", each colour is believed by Feng Shui practitioners to influence the wearer in different and positive ways; a combination of ruby red, emerald green, and amber yellow is most sought after for this purpose, the colours being said to enhance growth, wisdom, and wealth, respectively.

Ammolite is usually fashioned into freeform cabochons and mounted in gold, with diamonds as accents. Due to its delicacy, ammolite is best reserved for use in pendants, earrings, and brooches; if used as a ring stone, ammolite should be given a hard protective cap, namely one of synthetic spinel as used in triplets. Whole polished ammonites of appropriately small size may also be mounted in jewellery. Nothing harsher than mild soap and warm water should be used to clean ammolite jewellery; ultrasonic cleaning should be avoided.

Japan is the largest market for ammolite; this may be due to its use as an imitation of increasingly scarce black opal, or its aforementioned use in Feng Shui. Secondary markets include Canada, where it is used both by artisans who sell their creations to tourists of Banff National Park and in fine jewellery production; and the Southwest United States, where it is used by Zuni and other Native American craftspeople.

Grunerite

Grunerite is a mineral of the amphibole group of minerals with formula $Fe_7Si_8O_{22}(OH)_2$. It is the iron endmember of the grunerite-cummingtonite series. It forms as fibrous, columnar or massive aggregates of crystals. The crystals are monoclinic prismatic. The luster is glassy to pearly with colours ranging from green, brown to dark grey. The Mohs hardness is 5 to 6 and the specific gravity is 3.4 to 3.5.

It was discovered in 1853 and named for Louis Gruner, a Swiss-French chemist who first analysed it. Grunerite is also known commercially by the name *amosite* from the name Asbestos Mines Of South Africa (AMOSA).

Anyolite

Although anyolite is advertised as a variety of the mineral zoisite from Kenya and Tanzania, anyolite is actually a metamorphic rock composed of intergrown green zoisite, black tschermakite, and ruby. It is said to be named after the Maasai word *anyoli*, meaning “green.” Anyolite is also referred to as ruby in zoisite or Tanganyika artstone.

The contrasting colours make anyolite a popular material for sculptures and other decorative objects. It was first discovered at the Mundarara Mine, near Longido, Tanzania in 1954.

In 2010 it was suggested that a 2 kilogram stone known as the Gem of Tanzania owned by the defunct company Wrekin Construction and fraudulently valued at £11 million was actually a lump of Anyolite worth about £100, although it was eventually sold for £8000.

Antozonite

Antozonite (historically known as Stinkspat, Stinkfluss, Stinkstein, and fetid fluorite) is a radioactive fluorite variety first found in Wölsendorf, Bavaria, in 1841, and named in 1862.

It is characterized by the presence of multiple inclusions containing elemental fluorine; when the crystals are crushed or broken, the elemental fluorine is released. The fluorine subsequently reacts with atmospheric oxygen and water vapor, producing ozone (whose characteristic smell, originally mistaken for a hypothetical substance called antozone, is responsible for the mineral's name) and hydrogen fluoride.

Beryl

The mineral beryl is a beryllium aluminium cyclosilicate with the chemical formula $Be_3Al_2(SiO_3)_6$. The hexagonal crystals of beryl may be very small or range to several meters in size. Terminated crystals are relatively rare. Pure beryl is colourless, but it is frequently tinted by impurities; possible colours are green, blue, yellow, red, and white.

Etymology

The name beryl is derived from Greek *beryllos* which referred to a "precious blue-green colour-of-sea-water stone" and originated from Prakrit veruliya and Pali veluriya; veliru or, vilar, "to become pale"; ultimately from Sanskrit vaidurya-, which is of Dravidian origin, maybe from the name of Belur. The term was later adopted for the mineral beryl more exclusively. The Late Latin word *berillus* was abbreviated as *brill*-which produced the Italian word *brillare* meaning "shine", the French word *brille* meaning "shine" and the English word *brilliance.*

Deposits

Beryl of various colours is found most commonly in granitic pegmatites, but also occurs in mica schists in the Ural Mountains, and limestone in Colombia. Beryl is often associated with tin and tungsten ore bodies. Beryl is found in Europe in Norway, Austria, Germany, Sweden (especially morganite), Ireland and Russia, as well as Brazil, Colombia, Madagascar, Mozambique, South Africa, the United States, and Zambia. U.S. beryl locations are in California, Colorado, Connecticut, Idaho, Maine, New Hampshire, North Carolina, South Dakota and Utah.

New England's pegmatites have produced some of the largest beryls found, including one massive crystal from the Bumpus Quarry in Albany, Maine with dimensions 5.5 m by 1.2 m (18 ft by 4 ft) with a mass of around 18 metric tons; it is New Hampshire's state mineral. As of 1999, the largest known crystal of any mineral in the world is a crystal of beryl from Madagascar, 18 meters long and 3.5 meters in diameter.

Varieties

Aquamarine and Maxixe

Aquamarine is a blue or turquoise variety of beryl. It occurs at most localities which yield ordinary beryl, some of the finest coming from Russia. The gem-gravel placer deposits of Sri Lanka contain aquamarine. Clear yellow beryl, such as that occurring in Brazil, is sometimes called aquamarine chrysolite. When corundum presents the bluish tint of typical aquamarine, it is often termed Oriental aquamarine. The deep blue version of aquamarine is called *maxixe*. Its colour fades to white when exposed to sunlight or is subjected to heat treatment, though the colour returns with irradiation.

The pale blue colour of aquamarine is attributed to Fe^{2+}. The Fe^{3+} ions produce golden-yellow colour, and when both Fe^{2+} and Fe^{3+} are present, the colour is a darker blue as in maxixe. Decoloration of maxixe by light or heat thus may be due to the charge transfer Fe^{3+} and Fe^{2+}. Dark-blue maxixe colour can be produced in green, pink or yellow beryl by irradiating it with high-energy particles (gamma rays, neutrons or even X-rays).

In the United States, aquamarines can be found at the summit of Mt. Antero in the Sawatch Range in central Colorado. In Wyoming, aquamarine has been discovered in the Big Horn Mountains, near

Powder River Pass. In Brazil, there are mines in the states of Minas Gerais, Espírito Santo, and Bahia, and minorly in Rio Grande do Norte. The Mines of Colombia, Zambia, Madagascar, Malawi, Tanzania and Kenya also produce aquamarine.

The largest aquamarine of gemstone quality ever mined was found in Marambaia, Minas Gerais, Brazil, in 1910. It weighed over 110 kg, and its dimensions were 48.5 cm (19 in) long and 42 cm (17 in) in diameter.

Emerald

Emerald refers to green beryl, coloured by trace amounts of chromium and sometimes vanadium. The word "emerald" comes from Latin *smaragdus* from Greek *smaragdos* "green gem", its original source being a Semitic word *izmargad* or the Sanskrit word, *marakata*, meaning "green". Most emeralds are highly included, so their brittleness (resistance to breakage) is classified as generally poor.

Emeralds in antiquity were mined by the Egyptians and in Austria, as well as Swat in northern Pakistan. A rare type of emerald known as a trapiche emerald is occasionally found in the mines of Colombia. A trapiche emerald exhibits a "star" pattern; it has raylike spokes of dark carbon impurities that give the emerald a six-pointed radial pattern.

It is named for the *trapiche*, a grinding wheel used to process sugarcane in the region. Colombian emeralds are generally the most prized due to their transparency and fire. Some of the most rare emeralds come from three main emerald mining areas in Colombia: Muzo, Coscuez, and Chivor. Fine emeralds are also found in other countries, such as Zambia, Brazil, Zimbabwe, Madagascar, Pakistan, India, Afghanistan and Russia. In the US, emeralds can be found in Hiddenite, North Carolina. In 1998, emeralds were discovered in the Yukon.

Emerald is a rare and valuable gemstone and, as such, it has provided the incentive for developing synthetic emeralds. Both hydrothermal and *flux-growth* synthetics have been produced. The first commercially successful emerald synthesis process was that of Carroll Chatham. The other large producer of flux emeralds was Pierre Gilson Sr., which has been on the market since 1964. Gilson's emeralds are usually grown on natural colourless beryl seeds which become coated on both sides. Growth occurs at the rate of 1 mm per month, a typical seven-month growth run producing emerald crystals of 7 mm of thickness. The green colour of emeralds is attributed to presence of Cr^{3+} ions.

Golden Beryl and Heliodor

Golden beryl can range in colours from pale yellow to a brilliant gold. Unlike emerald, golden beryl has very few flaws. The term "golden beryl" is sometimes synonymous with *heliodor* but golden beryl refers to pure yellow or golden yellow shades, while heliodor refers to the greenish-yellow shades. The golden yellow colour is attributed to Fe^{3+} ions. Both golden beryl and heliodor are used as gems. Probably the largest cut golden beryl is the flawless 2054 carat stone on display in the Hall of Gems, Washington, D.C.

Goshenite

Colourless beryl is called *goshenite.* The name originates from Goshen, Massachusetts where it was originally discovered. Since all these colour varieties are caused by impurities and pure beryl is colourless, it might be tempting to assume that goshenite is the purest variety of beryl. However, there are several elements that can act as inhibitors to colour in beryl and so this assumption may not always be true. The name goshenite has been said to be on its way to extinction and yet it is still commonly used in the gemstone markets. Goshenite is found to some extent in almost all beryl localities. In the past, goshenite was used for manufacturing eyeglasses and lenses owing to its transparency. Nowadays, it is most commonly used for gemstone purposes and also considered as a source of beryllium.

The gem value of goshenite is relatively low. However, goshenite can be coloured yellow, green, pink, blue and in intermediate colours by irradiating it with high-energy particles. The resulting colour depends on the content of Ca, Sc, Ti, V, Fe, and Co impurities.

Morganite

Morganite, also known as "pink beryl", "rose beryl", "pink emerald", and "cesian (or *caesian*) beryl", is a rare light pink to rose-coloured gem-quality variety of beryl. Orange/yellow varieties of morganite can also be found, and colour banding is common. It can be routinely heat treated to remove patches of yellow and is occasionally treated by irradiation to improve its colour. The pink colour of morganite is attributed to Mn^{2+} ions.

Pink beryl of fine colour and good sizes was first discovered on an island on the coast of Madagascar in 1910. It was also known, with other gemstone minerals, such as tourmaline and kunzite, at Pala, California. In December 1910, the New York Academy of Sciences named the pink variety of beryl "morganite" after financier J. P. Morgan.

On October 7, 1989, one of the largest gem morganite specimens ever uncovered, eventually called "The Rose of Maine," was found at the Bennett Quarry in Buckfield, Maine, USA. The crystal, originally somewhat orange in hue, was 23 cm (9 in) long and about 30 cm (12 in) across, and weighed (along with its matrix) just over 50 lbs (23 kg).

Red Beryl

Red beryl (also known as "red emerald" or "scarlet emerald") is a red variety of beryl. It was first described in 1904 for an occurrence, its type locality, at Maynard's Claim (Pismire Knolls), Thomas Range, Juab County, Utah. The old synonym "bixbite" is deprecated from the CIBJO, because of the risk of confusion with the mineral bixbyite (also named after the mineralogist Maynard Bixby). The dark red colour is attributed to Mn^{3+} ions.

Red beryl is very rare and has only been reported from a handful of locations including: Wah Wah Mountains, Beaver County, Utah; Paramount Canyon, Sierra County, New Mexico; Round Mountain, Sierra County, New Mexico; and Juab County, Utah. The greatest concentration of gem-grade red beryl comes from the Violet Claim in the Wah Wah Mountains of mid-western Utah, discovered in 1958 by Lamar Hodges, of Fillmore, Utah, while he was prospecting for uranium. Prices for top quality natural red beryl can be as high as $10,000 per carat for faceted stones. Red beryl has been known to be confused with pezzottaite, also known as raspberry beryl or "raspberyl", a gemstone that has been found in Madagascar and now Afghanistan – although cut gems of the two varieties can be distinguished from their difference in refractive index.

While gem beryls are ordinarily found in pegmatites and certain metamorphic rocks, red beryl occurs in topaz-bearing rhyolites. It formed by crystallizing under low pressure and high temperature from a pneumatolitic phase along fractures or within near-surface miarolitic cavities of the rhyolite. Associated minerals include bixbyite, quartz, orthoclase, topaz, spessartine, pseudobrookite and hematite.

Argentite

Figure 4 : Argentite Sample

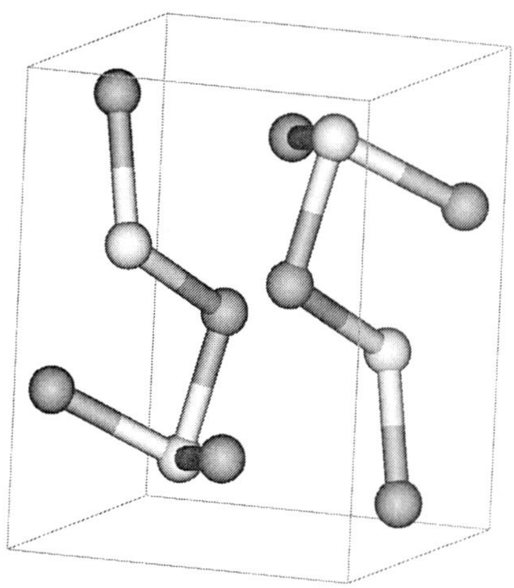

Figure 5 : The Unit cell of Argentite

Argentite was primarily treated as a mineral belonging to the galena group, cubic silver sulfide (Ag_2S), occasionally found as uneven cubes and octahedra, but more often as dendritic or earthy masses, with a blackish lead-grey colour and metallic lustre. All these forms are structurally not argentite, but its monoclinic polymorph, acanthite.

Due to instability in normal air temperature, the IMA commission decided to reject this mineral, which now is a discredited mineral species. The cubic cleavage, which is so prominent a feature in galena, here present only in traces. The mineral is perfectly sectile and has a shining streak; hardness 2.5, specific gravity 7.2-7.4. It occurs in mineral veins, and when found in large masses, as in is Mexico and in the Comstock Lode in Nevada, it forms an important ore of silver. The mineral was mentioned 1529 by G. Agricola, but the name argentite (from the Lat. *argentum*, silver) was not used till 1845 and is due to W. Haidinger. Old names for the species are Glaserz, silver-glance and vitreous silver. A related Cu-rich mineral occurring i.e. in Jalpa, Mexico, is known as jalpaite.

Asbestos

Figure 6 : Fibrous Asbestos on Muscovite

Figure 7 : Asbestos

Figure 8 : Asbestos

Asbestos is a set of six naturally occurring silicate minerals used commercially for their desirable physical properties. They all have in common their eponymous, asbestiform habit, long, (1:20) thin fibrous crystals.

The inhalation of asbestos fibres can cause serious illnesses, including malignant lung cancer, mesothelioma (a formerly rare cancer strongly associated with exposure to amphibole asbestos), and asbestosis (a type of pneumoconiosis). Long exposure to high concentrations of asbestos fibres is more likely to cause health problems. This is most common among the miners of asbestos, since they have the longest exposure to it. The European Union has banned all use of asbestos and extraction, manufacture and processing of asbestos products.

Asbestos became increasingly popular among manufacturers and builders in the late 19th century because of its sound absorption, average tensile strength, and its resistance to fire, heat, electrical and chemical damage. It was used in such applications as electrical insulation for hotplate wiring and in building insulation. When asbestos is used for its resistance to fire or heat, the fibres are often mixed with cement (resulting in fibre cement) or woven into fabric or mats. Commercial asbestos mining began in the Eastern Townships of Quebec, Canada and the world's largest asbestos mine is located in the town of Asbestos, Quebec.

Types and Associated Fibres

Figure 9 : Chrysotile Asbestos

Six minerals are defined by the United States Environmental Protection Agency as "asbestos" including those belonging to the serpentine class chrysotile and those belonging to the amphibole class amosite, crocidolite, tremolite, anthophyllite and actinolite. There is an important distinction to be made between serpentine and amphibole asbestos due to differences in their chemical composition and their degree of potency as a health hazard when inhaled. However asbestos and all commercial forms of asbestos (including chrysotile asbestos) are known to be human carcinogens based on sufficient evidence of carcinogenicity in humans.

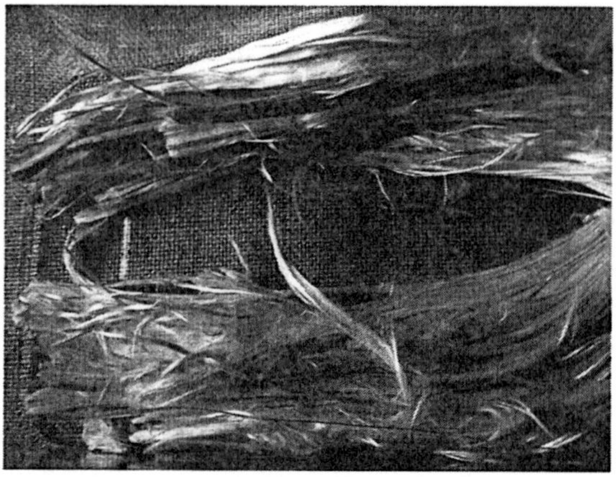

Figure 10 : Asbestos Fibres

Serpentine

White

Chrysotile, CAS No. 12001-29-5, is obtained from serpentinite rocks which are common throughout the world. Its idealized chemical formula is $Mg_3(Si_2O_5)(OH)_4$. Chrysotile fibres are curly as opposed to fibres from amosite, crocidolite, tremolite, actinolite, and anthophyllite which are needlelike. Chrysotile, along with other types of asbestos, has been banned in dozens of countries and is only allowed in the United States and Europe in very limited circumstances. Chrysotile has been used more than any other type and accounts for about 95% of the asbestos found in buildings in America. Applications where chrysotile might be used include the use of joint compound. It is more flexible than amphibole types of asbestos; it can be spun and woven into fabric. The most common use is within corrugated asbestos cement roof sheets typically used for outbuildings, warehouses and garages. It is also found as flat sheets used for ceilings and sometimes for walls and floors. Numerous other items have been made containing chrysotile including brake linings, cloth behind fuses (for fire protection), pipe insulation, floor tiles, and rope seals for boilers.

Amphibole

Brown

Amosite, CAS No. 12172-73-5, is a trade name for the amphiboles belonging to the cummingtonite-grunerite solid solution series, commonly from South Africa, named as an acronym from Asbestos Mines of South Africa. One formula given for amosite is $Fe_7Si_8O_{22}(OH)_2$. It is found most frequently as a fire retardant in thermal insulation products and ceiling tiles.

Blue

Crocidolite, CAS No. 12001-28-4 is an amphibole found primarily in southern Africa, but also in Australia. It is the fibrous form of the amphibole riebeckite. One formula given for crocidolite is $Na_2Fe^{2+}{}_3Fe^{3+}{}_2Si_8O_{22}(OH)_2$.

Crocidolite commonly occurs as soft friable fibres. Asbestiform amphibole may also occur as soft friable fibres but some varieties such as amosite are commonly straighter. All forms of asbestos are fibrillar in that they are composed of fibres with breadths less than 1 micrometer that occur in bundles and have very great widths. Asbestos with particularly fine fibres is also referred to as "amianthus". Amphiboles such as tremolite have a crystal structure containing strongly bonded ribbonlike silicate anion polymers that extend the width of the crystal. Serpentine (chrysotile) has a sheetlike silicate anion which is bowed and which rolls up like a carpet to form the fibre.

Other Materials

Other regulated asbestos minerals, such as tremolite asbestos, CAS No. 77536-68-6, $Ca_2Mg_5Si_8O_{22}(OH)_2$; actinolite asbestos, CAS No. 77536-66-4, $Ca_2(Mg, Fe)_5(Si_8O_{22})(OH)_2$; and anthophyllite asbestos, CAS No. 77536-67-5, $(Mg, Fe)_7Si_8O_{22}(OH)_2$; are less commonly used industrially but can still be found in a variety of construction materials and insulation materials and have been reported in the past to occur in a few consumer products.

Other natural and not currently regulated asbestiform minerals, such as richterite, $Na(CaNa)(Mg, Fe^{++})_5(Si_8O_{22})(OH)_2$, and winchite, $(CaNa)Mg_4(Al, Fe^{3+})(Si_8O_{22})(OH)_2$, may be found as a contaminant in products such as the vermiculite containing zonolite insulation manufactured by W.R. Grace and Company. These minerals are thought to be no less harmful than tremolite, amosite, or crocidolite, but since they are not regulated, they are referred to as "asbestiform" rather than asbestos although may still be related to diseases and hazardous.

Producing Nations

In 2009, 2 million tonnes of asbestos were mined worldwide. Russia was the largest producer with about 50% world share followed by China (14%), Brazil (12.5%), Kazakhstan (10.5%) and Canada (9%).

History

Early Uses

Asbestos use in human culture dates back at least 4,500 years, when evidence shows that inhabitants of the Lake Juojärvi region in East Finland strengthened earthenware pots and cooking utensils with the asbestos mineral anthophyllite. Asbestos was named by the ancient Greeks. One of the first careful descriptions of the material is attributed to Theophrastus in his text *On Stones*, around 300 BC, although the naming of minerals was not very consistent at that time. The term asbestos is traceable to Roman naturalist Pliny the Elder's manuscript *Natural History*, and his use of the term *asbestinon*, meaning "unquenchable". While Pliny is popularly attributed with recognising the detrimental effects of asbestos on slaves, examination of primary sources shows that this is not so. Charlemagne, the first Holy Roman Emperor (800-814), is said to have had a tablecloth made of asbestos.

According to Tabari, one of the curious items belonging to Khosrow II Parviz, the great Sassanian king (r. 531-579), was a napkin that he cleaned by simply throwing it into fire. This is believed to be made of asbestos.. Wealthy Persians, who bought asbestos imported over the Hindu Kush, amazed guests by cleaning the cloth by simply exposing it to fire. According to Biruni in his book of *Gems*, any cloths made of asbestos were called *shostakeh*. Some of the Persians believed the fibre was fur from an animal that lived in fire and died when exposed to water, hence the old mistaken myth that the salamander tolerated fire.

While travelling to Siberia, Marco Polo described being offered garments that could not burn. He was told that the wool was from the salamander, but did not accept this explanation. At last he was told that these garments were made from a mineral from the mountains, which contained threads just like wool.

Some archeologists believe that ancients made shrouds of asbestos, wherein they burned the bodies of their kings, in order to preserve only their ashes, and prevent their being mixed with those of wood or other combustible materials commonly used in funeral pyres. Others assert that the ancients used asbestos to make perpetual wicks for sepulchral or other lamps. In more recent centuries, asbestos was indeed used for this purpose. Although asbestos causes skin to itch upon contact, ancient literature indicates that it was prescribed for diseases of the skin, and particularly for the itch. It is possible that they used the term *asbestos* for soapstone, because the two terms have often been confused throughout history.

Industrial Era

The U.S. asbestos industry began in 1858 when fibrous anthophyllite was mined for use as asbestos insulation by the Johns Company, a predecessor to the current Johns Manville at a quarry at Ward's Hill on Staten Island, New York. Asbestos became more widespread during the industrial revolution; in 1866 it was used as insulation in the U.S. and Canada. Development of the first commercial asbestos mine began in 1874 in the Appalachian foothills of Quebec. By the mid 20th century uses included fire retardant coatings, concrete, bricks, pipes and fireplace cement, heat, fire, and acid resistant gaskets, pipe insulation, ceiling insulation, fireproof drywall, flooring, roofing, lawn furniture, and drywall joint compound.

Asbestos fibres were once used in automobile brake pads, shoes, and clutch discs. Since the mid-1990s, a majority of brake pads, new or replacement, have been manufactured instead with linings made of ceramic, carbon, metallic and aramid fibre (Twaron or Kevlar—the same material used in bulletproof vests).

Kent's filtered cigarette used crocidolite asbestos in its "Micronite" filter from 1952 to 1956.

Artificial Christmas snow, known as flocking, was previously made with asbestos.

In Japan, particularly after World War II, asbestos was used in the manufacture of ammonium sulfate for purposes of rice production, sprayed upon the ceilings, iron skeletons, and walls of railroad cars and buildings (during the 1960s), and used for energy efficiency reasons as well. Production of asbestos in Japan peaked in 1974 and went through ups and downs until about 1990, when production began to drop severely.

Discovery of Toxic Nature

The first documented death related to asbestos was in 1906. In the early 1900s researchers began to notice a large number of early deaths and lung problems in asbestos mining towns. The first diagnosis of asbestosis was made in the UK in 1924. By the 1930s, the UK regulated ventilation and made asbestosis an excusable work related disease, about ten years sooner than the U.S. The term mesothelioma was first used in medical literature in 1931; its association with asbestos was first noted sometime in the 1940s.

Approximately 100,000 people in the United States have died, or will die, from asbestos exposure related to ship building. In the Hampton Roads area, a shipbuilding centre, mesothelioma occurrence is seven times the national rate. Thousands of tons of asbestos were used in World War II ships to wrap the pipes, line the boilers, and cover engine and turbine parts. There were approximately 4.3 million shipyard workers in the United States during WWII; for every thousand workers about fourteen died of mesothelioma and an unknown number died from asbestosis.

The United States government and asbestos industry have been criticized for not acting quickly enough to inform the public of dangers, and to reduce public exposure. In the late 1970s court documents proved that asbestos industry officials knew of asbestos dangers since the 1930s and had concealed them from the public. A similar situation had arisen in the 1920s with the careless handling of radium and the ensuing scandal of the Radium Girls.

In Australia, asbestos was widely used in construction and other industries between 1945 and 1980. From the 1970s there was increasing concern about the dangers of asbestos and its use was phased out. Mining ceased in 1983. The use of asbestos was phased out in 1989 and banned entirely in Dec 2003. The dangers of asbestos are now well known in Australia and there is help and support for sufferers from asbestosis or mesothelioma.

Specific Products

Serpentine Group

Serpentine minerals have a sheet or layered structure. Chrysotile is the only asbestos mineral in the serpentine group. In the United States, chrysotile has been the most commonly used type of asbestos. According to the U.S. EPA Asbestos Building Inspectors Manual, chrysotile accounts for approximately 95% of asbestos found in buildings in the United States. Chrysotile is often present in a wide variety of products and materials, including:

- drywall and joint compound
- plaster
- gas mask filters pre 1960s

- mud and texture coats
- vinyl floor tiles, sheeting, adhesives
- roofing tars, felts, siding, and shingles
- "transite" panels, siding, countertops, and pipes
- popcorn ceilings, also known as acoustic ceilings
- fireproofing
- caulk
- gaskets
- packing, a system for sealing a rotating shaft
- brake pads and shoes
- clutch plates
- stage curtains
- fire blankets
- interior fire doors
- fireproof clothing for firefighters
- thermal pipe insulation
- filters for removing fine particulates from chemicals, liquids and wine
- dental cast linings
- HVAC flexible duct connectors
- drilling fluid additives.

In the European Union and Australia it has recently been banned as a potential health hazard and is not used at all. Japan is moving in the same direction, but more slowly. Revelations that hundreds of workers had died in Japan over the previous few decades from diseases related to asbestos sparked a scandal in mid-2005. Tokyo had, in 1971, ordered companies handling asbestos to install ventilators and check health on a regular basis; however, the Japanese government did not ban crocidolite and amosite until 1995, and a full-fledged ban on asbestos was implemented in October 2004.

Amphibole Group

Five types of asbestos are found in the amphibole group: amosite, crocidolite, anthophyllite, tremolite, and actinolite. Amosite, the second most likely type to be found in buildings, according to the U.S. EPA Asbestos Building Inspectors Guide, is the "brown" asbestos.

Amosite and crocidolite were formerly used in many products until the early 1980s. The use of all types of asbestos in the amphibole group was banned in much of the Western world by the mid-1980s, and by Japan in 1995. These products were mainly

- Low density insulation board and ceiling tiles;
- Asbestos-cement sheets and pipes for construction, casing for water and electrical/telecommunication services;
- Thermal and chemical insulation (*e.g., fire rated doors, limpet spray, lagging and gaskets*).

Health Problems

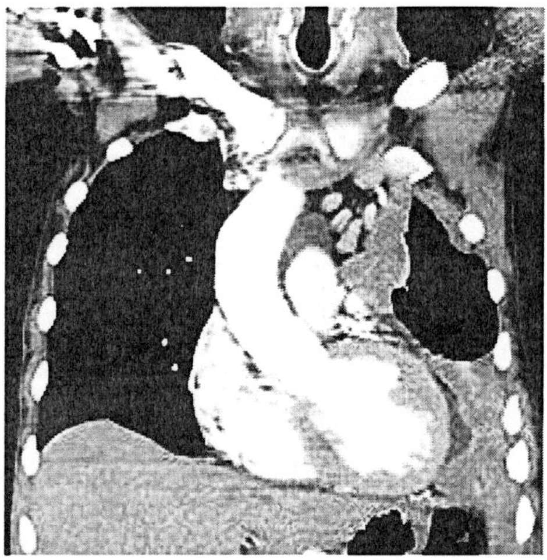

Figure 11: Left-sided mesothelioma: chest CT

Amosite and crocidolite are the most hazardous of the asbestos minerals because of their long persistence in the lungs of exposed people. Tremolite often contaminates chrysotile asbestos, thus creating an additional hazard.

Chrysotile asbestos, like all other forms of asbestos, has produced tumors in animals. Mesotheliomas have been observed in people who were occupationally exposed to chrysotile, family members of the occupationally exposed, and residents who lived close to asbestos factories and mines.

According to the NCI, "A history of asbestos exposure at work is reported in about 70 percent to 80 percent of all cases. However, mesothelioma has been reported in some individuals without any known exposure to asbestos." The most common diseases associated with chronic exposure to asbestos include: asbestosis and pleural abnormalities (mesothelioma, lung cancer). Asbestos exposure becomes a health concern when high concentrations of asbestos fibres are inhaled over a long time period. People who become ill from inhaling asbestos are often those who are exposed on a day-to-day basis in a job where they worked directly with the material.

As a person's exposure to fibres increases, because of being exposed to higher concentrations of fibres and/or by being exposed for a longer time, then that person's risk of disease also increases. Disease is very unlikely to result from a single, high-level exposure, or from a short period of exposure to lower levels.

Mechanisms Which Might be Triggering Cancer Development

Stanton and Layard concluded in 1977–78 that asbestos toxicity is *not* initiated by *chemical* effects, that is any trigger-effects of asbestos must presumably be *physical*, such as (A) *mechanical* damage or (B) unwanted *signal channels* (a plausible property for slender transparent fibres) which might disrupt normal cell activity—especially mitosis.

(A) Mechanical Damage. There is experimental evidence that very slim fibres (<60 nm, <0.06 μm in breadth) do tangle destructively with chromosomes (being of comparable size). Clearly that is likely to cause the sort of mitosis disruption expected in cancer.

(B) Unwanted Signal channels. This has recently been explored theoretically, but not yet experimentally. The theory argues that this effect would only be feasible for asbestos fibres >120 nm in breadth, which suggests that we should be on the look-out for a possible mixture of different mechanisms for the different fibre-diameter-ranges.

One popular idea of the causal chain is (1) Asbestos fibre →→ (3) inflammation →(4) other pathology. While that may be true, it does not explain "(2), the actual trigger":

"What is the *physical* property of asbestos which *initiates* any such inflammation?" (After all, inflammation is usually seen as caused by *chemical*-based processes: immunological &/or bacterial). So inflammation (&/or oxidation etc.) may well be part of the causal chain, *but not the crucial first step*.

Other asbestos-related Diseases

- Asbestosis: Progressive fibrosis of the lungs of varying severity, progressing to bilateral fibrosis, honeycombing of the lungs on radiological view with symptoms including rales and wheezing.

- Asbestos warts: caused when the sharp fibres lodge in the skin and are overgrown causing benign callus-like growths.
- Pleural plaques: discrete fibrous or partially calcified thickened area which can be seen on X-rays of individuals exposed to asbestos. Although pleural plaques are themselves asymptomatic, in some patients this develops into pleural thickening.
- Diffuse pleural thickening: similar to above and can sometimes be associated with asbestosis. Usually no symptoms shown but if exposure is extensive, it can cause lung impairment.

Asbestos as a Contaminant

Most respirable asbestos fibres are invisible to the unaided human eye because their size is about 3–20 µm wide and can be as slim as 0.01 µm. Human hair ranges in size from 17 to 181 µm in breadth. Fibres ultimately form because when these minerals originally cooled and crystallized, they formed by the polymeric molecules lining up parallel with each other and forming oriented crystal lattices. These crystals thus have three cleavage planes, and in this case, there are two cleavage planes which are much weaker than the third. When sufficient force is applied, they tend to break along their weakest directions, resulting in a linear fragmentation pattern and hence a fibrous form. This fracture process can keep occurring and one larger asbestos fibre can ultimately become the source of hundreds of much thinner and smaller fibres.

As asbestos fibres get smaller and lighter, they more easily become airborne and human respiratory exposures can result. Fibres will eventually settle but may be re-suspended by air currents or other movement.

Friability of a product containing asbestos means that it is so soft and weak in structure that it can be broken with simple finger crushing pressure. Friable materials are of the most initial concern because of their ease of damage. The forces or conditions of usage that come into intimate contact with most non-friable materials containing asbestos are substantially higher than finger pressure.

Environmental Asbestos

Asbestos can be found naturally in the air outdoors and in some drinkable water, including water from natural sources. Studies have shown that members of the general (non-occupationally exposed) population have tens of thousands to hundreds of thousands of asbestos fibres in each gram of dry lung tissue, which translates into millions of fibres and tens of thousands of asbestos bodies in every person's lungs.

Asbestos from natural geologic deposits is known as "naturally occurring asbestos" (NOA). Health risks associated with exposure to NOA are not yet fully understood, and current US federal regulations do not address exposure from NOA. Many populated areas are in proximity to shallow, natural deposits which occur in 50 of 58 California counties and in 19 other U.S. states. In one study, data was collected from 3,000 mesothelioma patients in California and 890 men with prostate cancer, a malignancy not known to be related to asbestos. The study found a correlation between the incidence of mesotheliomas and the distance a patient lived from known deposits of rock likely to include asbestos; the correlation was not present when the incidence of prostate cancer was compared with the same distances. According to the study, risk of mesothelioma declined by 6% for every 10 km that an individual had lived away from a likely asbestos source.

Portions of El Dorado County, California are known to contain natural amphibole asbestos formations at the surface. The USGS studied amphiboles in rock and soil in the area in response to an EPA sampling study and subsequent criticism of the EPA study. The EPA study was refuted by its own peer reviewers and never completed or published. The study found that many amphibole particles in the area meet the counting rule criteria used by the EPA for chemical and morphological limits, but do not meet morphological requirements for commercial-grade-asbestos. The executive summary pointed out that even particles that do not meet requirements for commercial-grade-asbestos may be a health threat and suggested a collaborative research effort to assess health risks associated with "Naturally Occurring Asbestos."

However, the main criticism pointed at EPA was that their testing was conducted in small isolated areas of El Dorado where there were no amphibole asbestos deposits, thus the language regarding amphibole, nonfibrous "particles". Actual surface amphibole deposits in residential areas were ignored for testing purposes. Thus no final findings were published by ATSDR since the criticism was correct and the effort of combined EPA/ATSDR teams were wasted time and money.

Great deals of Fairfax County, Virginia were also found to be underlain with tremolite. The county monitored air quality at construction sites, controlled soil taken from affected areas, and required freshly developed sites to lay 6 inches (150 mm) of clean, stable material over the ground.

For environmental samples, one must normally resort to electron microscopy for positive identification. Today, gravimetric and PCM/PLM techniques are employed. However, the latter techniques cannot readily identify the smallest, most hazardous, fibres, because they are limited to PM10 particulate size evaluation, which completely ignores ultrafine particles (UFPs).

History of Health Concerns and Regulation

Until 1900

By the first century AD, Greeks and Romans are claimed to have observed that slaves involved in the weaving of asbestos cloth were afflicted with a sickness of the lungs, although this is not confirmed by examination of primary sources.

Early concern in the modern era on the health effects of asbestos exposure can be found in several sources. Among the earliest were reports in Britain. The annual reports of the Chief Inspector of Factories reported as early as 1898 that asbestos had "easily demonstrated" health risks.

At about the same time, what was probably the first study of mortality among asbestos workers was reported in France. While the study describes the cause of death as chalicosis, a generalized pneumoconiosis, the circumstances of the employment of the fifty workers whose death prompted the study suggest that the root cause was asbestos or mixed asbestos-cotton dust exposure.

1900s 1910s

Further awareness of asbestos-related diseases can be found in the early 1900s, when London doctor H. Montague Murray conducted a post mortem exam on a young asbestos factory worker who died in 1899. Dr. Murray gave testimony on this death in connection with an industrial disease compensation hearing. The post-mortem confirmed the presence of asbestos in the lung tissue, prompting Dr. Murray to express as an expert opinion his belief that the inhalation of asbestos dust had at least contributed to, if not actually caused, the death of the worker.

The record in the United States was similar. Early observations were largely anecdotal in nature and did not definitively link the occupation with the disease, followed by more compelling and larger studies that strengthened the association. One such study, published in 1918, noted:

All of these processes unquestionably involve a considerable dust hazard, but the hygienic aspects of the industry have not been reported upon. It may be said, in conclusion, that in the practice of American and Canadian life insurance companies, asbestos workers are generally declined on account of the assumed health-injurious conditions of the industry.

1920s 1930s

Widespread recognition of the occupational risks of asbestos in Britain was reported in 1924 by a Dr. Cooke, a pathologist, who introduced a case description of a 33-year-old female asbestos worker, Nellie Kershaw, with the following: "Medical men in areas where asbestos is manufactured have long suspected the dust to be the cause of chronic bronchitis and fibrosis..." Dr. Cooke then went on to report on a case in 1927 involving a 33-year-old male worker who was the only survivor out of ten workers in an asbestos carding room. In the report he named the disease "asbestosis".

Dr. Cooke's second case report was followed, in the late 1920s, by a large public health investigation (now known as the Merewether report after one of its two authors) that examined some 360 asbestos-textile workers (reported to be about 15% of the total comparable employment in Britain at the time) and found that about a quarter of them suffered from pulmonary fibrosis. This investigation resulted in improved regulation of the manufacturing of asbestos-containing products in the early 1930s. Regulations included industrial hygiene standards, medical examinations, and inclusion of the asbestos industry into the British Workers' Compensation Act.

The first known U.S. workers' compensation claim for asbestos disease was in 1927. In 1930, the first reported autopsy of an asbestosis sufferer was conducted in the United states and later presented by a doctor at the Mayo Clinic, although in this case the exposure involved mining activities somewhere in South America.

In 1930, the major asbestos company Johns-Manville produced a report, for internal company use only, about medical reports of asbestos worker fatalities. In 1932, a letter from U.S. Bureau of Mines to asbestos manufacturer Eagle-Picher stated, in relevant part, "It is now known that asbestos dust is one of the most dangerous dusts to which man is exposed." In 1933, Metropolitan Life Insurance Co. doctors found that 29% of workers in a Johns-Manville plant had asbestosis. Likewise, in 1933, Johns-Manville officials settled lawsuits by 11 employees with asbestosis on the condition that the employees' lawyer agree to never again "directly or indirectly participate in the bringing of new actions against the Corporation."

In 1934, officials of two large asbestos companies, Johns-Manville and Raybestos-Manhattan, edited an article about the diseases of asbestos workers written by a Metropolitan Life Insurance Company doctor. The changes downplayed the danger of asbestos dust. In 1935, officials of Johns-Manville and Raybestos-Manhattan instructed the editor of *Asbestos* magazine to publish nothing about asbestosis. In 1936, a group of asbestos companies agreed to sponsor research on the health effects of asbestos dust, but required that the companies maintain complete control over the disclosure of the results.

1940s

In 1942, an internal Owens-Corning corporate memo referred to "medical literature on asbestosis... scores of publications in which the lung and skin hazards of asbestos are discussed." Testimony given in a federal court in 1984 by Charles H. Roemer, formerly an employee of Unarco, described a meeting in the early 1940s between Unarco officials, J-M President Lewis H. Brown and J-M attorney Vandiver Brown.

Roemer stated, "I'll never forget, I turned to Mr. Brown, one of the Browns made this crack (that Unarco managers were a bunch of fools for notifying employees who had asbestosis), and I said, 'Mr. Brown, do you mean to tell me you would let them work until they dropped dead?' He said, 'Yes. We save a lot of money that way.'" In 1944, a Metropolitan Life Insurance Company report found 42 cases of asbestosis among 195 asbestos miners.

1950s

In 1951, asbestos companies removed all references to cancer before allowing publication of research they sponsored. In 1952, Dr. Kenneth Smith, Johns-Manville medical director, recommended (unsuccessfully) that warning labels be attached to products containing asbestos. Later, Smith testified: "It was a business decision as far as I could understand... the corporation is in business to provide jobs for people and make money for stockholders and they had to take into consideration the effects of everything they did and if the application of a caution label identifying a product as hazardous would cut into sales, there would be serious financial implications." In 1953, National Gypsum's safety director wrote to the Indiana Division of Industrial Hygiene, recommending that acoustic plaster mixers wear respirators "because of the asbestos used in the product." Another company official noted that the letter was "full of dynamite" and urged that it be retrieved before reaching its destination. A memo in the files noted that the company "succeeded in stopping" the letter, which "will be modified."

1960s 1980s

Through the 1970s, asbestos was used to fireproof roofing and flooring, for heat insulation, and for a variety of other purposes. The material was used in fire-check partitioning and doors on North Sea Oil Production Platforms and Rigs.

Modern Regulation

United States

In 1989 the United States Environmental Protection Agency (EPA) issued the Asbestos Ban and Phase Out Rule which was subsequently overturned in the case of Corrosion Proof Fittings v. EPA, 947 F.2d 1201 (5th Cir. 1991). This ruling leaves many consumer products that can still legally contain trace amounts of asbestos. For a clarification of products which legally contain asbestos, read the EPA's clarification statement.

The EPA has proposed a concentration limit of seven million fibres per liter of drinking water for long fibres (lengths greater than or equal to 5 μm). The Occupational Safety and Health Administration (OSHA), has set limits of 100,000 fibres with lengths greater than or equal to 5 μm per cubic meter of workplace air for eight-hour shifts and 40-hour work weeks.

OSHA regulations regarding asbestos are covered in 29 C.F.R. 1926.1101. Such work is divided into four categories.

Class I asbestos work means activities involving the removal of TSI and surfacing ACM and PACM.

Class II asbestos work means activities involving the removal of ACM which is not thermal system insulation or surfacing material. This includes, but is not limited to, the removal of asbestos-containing wallboard, floor tile and sheeting, roofing and siding shingles, and construction mastics.

Class III asbestos work means repair and maintenance operations, where "ACM", including TSI and surfacing ACM and PACM, is likely to be disturbed.

Class IV asbestos work means maintenance and custodial activities during which employees contact but do not disturb ACM or PACM and activities to clean up dust, waste and debris resulting from Class I, II, and III activities.

New Zealand

In 1984, the import of raw amphibole (blue and brown) asbestos into New Zealand was banned. In 2002 the import of chrysotile (white) asbestos was banned.

Australia

A complete ban on asbestos-containing material in Australia was introduced in 1991 although some building materials in storage were still being used in the years that followed. Queensland began regulation of asbestos removal and disposal in 2005. Handlers of asbestos materials must have a B-Class license for bonded asbestos and an A-Class license for friable asbestos.

The town of Wittenoom, in Western Australia was built around a (blue) asbestos mine. The entire town continues to be contaminated, and has been degazetted, allowing local authorities to remove references to Wittenoom from maps and roadsigns.

Contamination of Other Products

Asbestos and Vermiculite

Vermiculite is a hydrated laminar magnesium-aluminum-iron silicate which resembles mica. It can be used for many industrial applications and has been used as a replacement for asbestos. Some ore bodies of vermiculite have been found to contain small amounts of asbestos. One vermiculite mine operated by W. R. Grace and Company in Libby, Montana exposed workers and community residents to danger by mining contaminated vermiculite. In 1999 the EPA began cleanup efforts and now the area is a Superfund cleanup area. The EPA has determined that harmful asbestos is released from the mine as well as through other activities that disturb soil in the area.

Asbestos and Talc

Talc is sometimes contaminated with asbestos. In 2000, tests in a certified asbestos-testing laboratory found the tremolite form of amphibole asbestos in three out of eight bigger brands of children's crayons that are made partly from talc: Crayola, Prang, and RoseArt. In Crayola crayons, the tests found asbestos levels from 0.05% in *Carnation Pink* to 2.86% in *Orchid*; in Prang crayons, the range was from 0.3% in *Periwinkle* to 0.54% in *Yellow*; in Rose Art crayons, it was from 0.03% in *Brown* to 1.20% in *Orange*.

Overall, 32 different types of crayons from these brands contained more than trace amounts of asbestos, and eight others contained trace amounts. The Art and Creative Materials Institute, a trade association which tests the safety of crayons on behalf of the makers, initially insisted the test results must be incorrect, although they later said they do not test for asbestos.

In May 2000, Crayola said tests by a materials analyst, Richard Lee, whose testimony has been accepted in lawsuits over 250 times on behalf of the asbestos industry, showed two of its crayons were negative for asbestos. In June 2000, Binney & Smith, the maker of Crayola, and the other makers agreed to stop using talc in their products, and changed their product formulations in the United States. The mining company, R T Vanderbilt Co of Gouverneur, New York, which supplied the talc to the crayon makers, insists there is no asbestos in its talc "to the best of our knowledge and belief", but a news article claimed that the United States Mine Safety and Health Administration (MSHA) did find asbestos in four talc samples that it tested in 2000.

At the time, however, the Assistant Secretary for Mine Safety and Health informed the news reporter that his article was in error and that the reporter had misquoted him stating that "In fact, the abbreviation ND (non detect) in the laboratory report – indicates no asbestos fibres actually were found in the samples." Further supporting the claim of Vanderbilt that asbestos is not found in this industrial grade talc (composed of a very complex mineral mixture) is a decades old record of analytical work that does not find asbestos in this talc by mineral scientists in academia, government and contract laboratories.

Human, animal and cell health studies conducted on Vanderbilt's controversial talc also lend no support for the presence of asbestos in this talc. Several non fully peer-reviewed health reports concerning Vanderbilt talc do exist and suggest a "same as" asbestos risk, some of which were referenced in the previously cited news articles.

Asbestos in Construction

Asbestos Construction in Developed Countries

The use of asbestos in new construction projects has been banned for health and safety reasons in many developed countries or regions, including the European Union, Australia, Hong Kong, Japan, and New Zealand. A notable exception is the United States, where asbestos continues to be used in construction such as cement asbestos pipes. The 5th Circuit Court prevented the EPA from banning asbestos in 1991 because although EPA research showed it would cost between $450 and 800 million and save around 200 lives in a 13-year length, the EPA did not provide adequate evidence for the safety of alternative products. Until the mid-1980s, small amounts of white asbestos were used in the manufacture of Artex, a decorative stipple finish, however, some of the lesser-known suppliers of Artex were still adding white asbestos until 1999. Removing or disturbing Artex is not recommended, as it may contain white asbestos.

Prior to the ban, asbestos was widely used in the construction industry. Many older buildings built before the late 1990s contain asbestos. In the United States, there is a minimum standard for asbestos surveys as described by ASTM Standard E 2356-04. The U.S. Environmental Protection Agency includes some but not all asbestos-contaminated facilities on the Superfund National Priorities list (NPL). Renovation and demolition of asbestos contaminated buildings is subject to EPA NESHAP and OSHA Regulations. Asbestos is not a material covered under CERCLA's innocent purchaser defence. In the UK, the removal and disposal of asbestos and of substances containing it are covered by the Control of Asbestos Regulations 2006

In older buildings (e.g. those built prior to 1999 in the UK, before white asbestos was finally banned), asbestos may still be present in some areas e.g. old bath panels, concrete water tanks and many other places. Being aware of asbestos locations reduces the risk of disturbing asbestos.

Removal of asbestos building components can also remove the fire protection they provide, therefore fire protection substitutes are required for proper fire protection that the asbestos originally provided.

Asbestos Construction in Developing Countries

Some developing countries, such as India and China, and also Russia, have continued widespread use of asbestos. The most common is corrugated asbestos-cement sheets or "A/C Sheets" for roofing and for side walls. Millions of homes, factories, schools or sheds and shelters continue to use asbestos. Cutting these sheets to size and drilling holes to receive 'J' bolts to help secure the sheets to roof framing is done on-site. There has been no significant change in production and use of A/C Sheets in developing countries following the widespread restrictions in developed nations.

Asbestos and 9/11

More than 1,000 tons of asbestos are thought to have been released into the air during the destruction of the Twin Towers in New York on 9/11. Inhalation of a mixture of asbestos and other toxicants is thought to be linked to the unusually high death rate of emergency service workers from cancer since the disaster. Many thousands more are now thought to be at risk of developing cancer due to this exposure with those who have died so far being only the 'tip of the iceberg'. Some commentators have criticised authorities for using asbestos in the Towers' construction.

Litigation

Asbestos litigation is the longest, most expensive mass tort in U.S. history, involving more than 8,400 defendants and 730,000 claimants as of 2002 according to the RAND Corporation, and at least one defendant reported claim counts in excess of 800,000 in 2006.

Current trends indicate that the worldwide rate at which people are diagnosed with asbestos-related diseases will likely increase through the next decade. Analysts have estimated that the total costs of asbestos litigation in the USA alone is over $250 billion.

The federal legal system in the United States has dealt with numerous counts of asbestos related suits, which often included multiple plaintiffs with similar symptoms. In 1999 there were 200,000 related cases pending in the federal court system of the United States. Further, it is estimated that within the next 40 years, the number of cases may increase to 700,000. These numbers help explain how there are thousands of current pending cases.

Litigation of asbestos materials has been slow. Companies sometimes counter saying that health issues do not currently appear in their worker or workers, or sometimes are settled out of court. The Research and Development (RAND) think tank has appropriated certain legal information which is readily available for proclaimed victims of natural resource accidents. This information has helped many workers, regardless of health condition, earn compensation through companies. RAND, along with the Institute for Civil Justice (ICJ) have been proponents of the organization of past cases in order to determine one aspect of fair compensation for workers.

1999 saw the introduction of the Fairness in Asbestos Compensation Act. Ultimately many asbestos companies were forced to file for bankruptcy. While companies filed for bankruptcy, this limited payouts to those who were actually affected by the material. Christopher Edley said what the 1999 Act ultimately did was "limit punitive damages that seek retribution for the decisions of long-dead executives for conduct that took place decades ago (Professor Christopher Edley, Jr.)."

Litigation exists outside the United States in England, Scotland, Ireland, the Netherlands, France, Australia, and Japan among other nations.

The volume of the asbestos liability has concerned manufacturers and insurers and reinsurers. The amounts and method of allocating compensation have been the source of many court cases, and government attempts at resolution of existing and future cases.

Critics of Safety Regulations

Canada-EU Dispute

Natural Resources Canada states chrysotile, one of the fibres that make up asbestos, is not as dangerous as once thought. According to their fact sheet, "current knowledge and modern technology can successfully control the potential for health and environmental harm posed by chrysotile".

In May 1998, Canada requested consultations before the WTO and the European Commission concerning France's 1996 prohibition of the importation and sale of asbestos. Canada said that the French measures contravened provisions of the Agreements on Sanitary and Phytosanitary Measures and on Technical Barriers to Trade, and the GATT 1994. The EC claims that substitute materials have been developed in place of asbestos, which are safer to human health. It stressed that the French measures were not discriminatory, and were fully justified for public health reasons. The EC claimed that in the July consultations, it had tried to convince Canada that the measures were justified, and that just as Canada broke off consultations, it was in the process of submitting substantial scientific data in favour of the asbestos ban.

The Canadian federal government has in response claimed that chrysotile is much less dangerous than other types of asbestos, and Canada does not export the other asbestos fibre. Chrysotile continues to be used in new construction across Canada, in ways that are very similar to those for which chrysotile is exported. The Chrysotile Institute, an asbestos industry funded organization, said that the use of chrysotile does not pose an environmental problem and the inherent risks in its use are limited to the workplace. The Canadian government continues to draw both domestic and international criticism for its stance on chrysotile, most recently in international meetings on the Rotterdam Convention hearings regarding chrysotile.

The CFMEU pointed out that most exports go to developing countries. Canada has pressured countries, including Chile, and other UN member states to avoid chrysotile bans.

Other Criticism

Asbestos regulation critics include the asbestos industry and Fox News "junk science" columnist Steven Milloy. Critics sometimes argue that increased government regulation does more harm than good and that replacements to asbestos are inferior. An example is the suggestion by Dixy Lee Ray and others that the shuttle Challenger disintegrated because the maker of O-ring putty was pressured by the EPA into ceasing production of asbestos-laden putty. However, the putty used in *Challenger*'s final flight *did* contain asbestos, and failures in the putty were not responsible for the failure of the O-ring that led to loss of the shuttle.

Asbestos was used in the first forty floors of the World Trade Centre north tower causing an airborne contamination among lower Manhattan after the towers collapsed in the attacks on September 11th, 2001. Steven Milloy suggests that the World Trade Centre towers could still be standing or at least would have stood longer had a 1971 ban not stopped the completion of the asbestos coating above the 64th floor. This was not considered in the National Institute of Standards and Technology's report on the towers' collapse. All fireproofing materials, regardless of what they are made of, are required to obtain a fire-resistance rating prior to installation. All fibre-based lightweight commercial spray fireproofing materials are vulnerable to kinetic energy impacts that are outside of the fire testing upon which their ratings are based, including asbestos-based materials, and may have been removed in large areas by the impact of the planes.

Substitutes for Asbestos in Construction

Fiberglass insulation was invented in 1938 and is now the most commonly used type of insulation material. The safety of this material is also being called into question, as research shows that the composition of this material (asbestos and fiberglass are both silicate fibres) causes similar toxicity as asbestos.

In 1978, a highly texturized fiberglass fabric was invented by Bal Dixit, called Zetex. This fabric is lighter than asbestos, but offers the same bulk, thickness, hand, feel, and abrasion resistance as asbestos. The fiberglass was texturized to eliminate some of the problems that arise with fiberglass, such as poor abrasion resistance and poor seam strength

In Europe stone-and glasswool are the main insulators in houses.

Many companies that produced asbestos-cement products that were reinforced with asbestos fibres have developed products incorporating organic fibres. One such product was known as Eternit and another "Everite" now use "Nutec" fibres which consist of organic fibres, portland cement and silica. Cement-bonded wood fibre is another substitute. Stone fibres are used in gaskets and friction materials.

Another potential fibre is polybenzimidazole or PBI fibre. Polybenzimidazole fibre is a synthetic fibre with high melting point of 760 °C that also does not ignite. Because of its exceptional thermal and chemical stability, it is often used by fire departments and space agencies.

Asbestos alternatives for industrial use include sleeves, rope, tape, fabric, textiles and insulation batt materials made from fiberglass and silica.

Recycling and Disposal

In most developed countries, asbestos is typically disposed of as hazardous waste in landfill sites.

Asbestos can also be recycled by transforming it into harmless silicate glass. A process of thermal decomposition at 1000–1250 °C produces a mixture of non-hazardous silicate phases, and at temperatures above 1250 °C it produces silicate glass. Microwave thermal treatment can be used in an industrial manufacturing process to transform asbestos and asbestos-containing waste into porcelain stoneware tiles, porous single-fired wall tiles, and ceramic bricks.

Illite

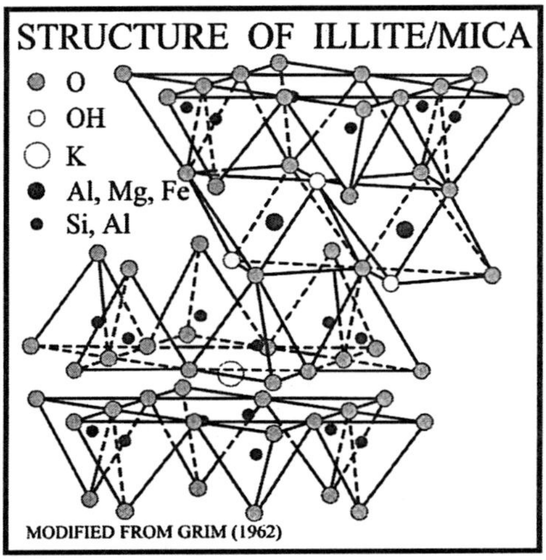

Figure 12 : Structure of Illite mica - USGS.

Illite is a non-expanding, clay-sized, micaceous mineral. Illite is a phyllosilicate or layered alumino-silicate. Its structure is constituted by the repetition of tetrahedron – octahedron – tetrahedron (TOT) layers. The interlayer space is mainly occupied by poorly hydrated potassium cations responsible for the absence of swelling. Structurally illite is quite similar to muscovite with slightly more silicon, magnesium, iron, and water and slightly less tetrahedral aluminium and interlayer potassium. The chemical formula is given as $(K,H_3O)(Al,Mg,Fe)_2(Si,Al)_4O_{10}[(OH)_2,(H_2O)]$, but there is considerable ion substitution. It occurs as aggregates of small monoclinic grey to white crystals. Due to the small size, positive identification usually requires x-ray diffraction or SEM-EDS (automated mineralogy) analysis. Illite occurs as an alteration product of muscovite and feldspar in weathering and hydrothermal environments; it may be a component of sericite. It is common in sediments, soils, and argillaceous sedimentary rocks as well as in some low grade metamorphic rocks. The iron rich member of the illite group, glauconite, in sediments can be differentiated by x-ray analysis.

The cation exchange capacity (CEC) of illite is smaller than that of smectite but higher than that of kaolinite, typically around 20 – 30 meq/100 g.

Illite was first described for occurrences in the Maquoketa shale in Calhoun County, Illinois, USA, in 1937. The name was derived from its type location in Illinois.

Illite is also called hydromica or hydromuscovite. Brammallite is a sodium rich analogue. Avalite is a chromium bearing variety which has been described form Mt. Avala, Belgrade, Serbia.

Illite Crystallinity

The crystallinity of illite has been used as an indicator of metamorphic grade in clay-bearing rocks metamorphosed under conditions between diagenesis and low-grade. With increasing temperature illite is thought to undergo a transformation into muscovite.

Aventurine

Aventurine is a form of quartz, characterised by its translucency and the presence of platy mineral inclusions that give a shimmering or glistening effect termed *aventurescence*.

Figure 13 : Aventurine (unknown scale)

The most common colour of aventurine is green, but it may also be orange, brown, yellow, blue, or gray. Chrome-bearing fuchsite (a variety of muscovite mica) is the classic inclusion, and gives a silvery green or blue sheen.

Oranges and browns are attributed to hematite or goethite. Because aventurine is a rock, its physical properties vary: its specific gravity may lie between 2.64-2.69 and its hardness is somewhat lower than single-crystal quartz at around 6.5.

Aventurine feldspar or sunstone can be confused with orange and red aventurine quartzite, although the former is generally of a higher transparency. Aventurine is often banded and an overabundance of fuchsite may render it opaque, in which case it may be mistaken for malachite at first glance.

The name *aventurine* derives from the Italian "a ventura" meaning "by chance". This is an allusion to the lucky discovery of aventurine glass or goldstone at some point in the 18th century. Although it was known first, goldstone is now a common imitation of aventurine and sunstone.

Goldstone is distinguished visually from the latter two minerals by its coarse flecks of copper, dispersed within the glass in an unnaturally uniform manner. It is usually a golden brown, but may also be found in blue or green.

Figure 14: Aventurine (unknown scale)

The majority of green and blue-green aventurine originates in India (particularly in the vicinity of Mysore and Madras) where it is employed by prolific artisans. Creamy white, gray and orange material is found in Chile, Spain and Russia. Most material is carved into beads and figurines with only the finer examples fashioned into cabochons, later being set into jewellery.

Aksaite

Aksaite ($Mg[B_6O_7(OH)_6]\cdot 2H_2O$) is a mineral found in Kazakhstan.

Etymology and History

Aksaite is named after the place it was discovered, *Ak-say* (*lit.* White Glen). It was found in the year 1963 in Chelkar Salt Dome, Ak-say Valley, Kazakhstan.

Alabandite

Alabandite or alabandine is a rarely occurring manganese sulfide mineral. It crystallizes in the cubic crystal system with the chemical composition $Mn^{2+}S$ and develops commonly massive to granular aggregates, but rarely also cubic or octahedral crystals to 1 cm.

Etymology and History

Alabandite was first described in 1784 by Franz-Joseph Müller von Reichenstein. The mineral name is derived from its supposed discovery locality at Alabanda (Aïdin) in Turkey.

Occurrence

Alabandite forms in epithermal polymetallic sulfide veins and low-temperature manganese deposits. It occurs with acanthite, calcite, chalcopyrite, galena, pyrite, quartz, rhodochrosite, rhodonite, sphalerite and native tellurium. Sometimes it was found in meteorites.

Localities are several areas in Antarctica, Argentina, Armenia, Australia, Bolivia, Brazil, Bulgaria, Canada, China, Czech Republic, Finland, France, Germany, Ghana, Greece, Greenland, India, Italy, Japan, Kyrgyzstan, Mexico, New Zealand, Norway, Peru, Poland, Romania, Russia, Slovakia, South Africa, South Korea, Sweden, Switzerland, Taiwan, Tanzania, the United Kingdom, the USA, Uzbekistan and Yemen. Altogether at present time approximately 220 discovery sites are registered.

Crystal structure

Alabandite crystallizes in the cubic crystal system in the space group Fm3m with the lattice parameter a = 5.22 Å and four formula units per unit cell.

Alamosite

Alamosite ($Pb_{12}Si_{12}O_{36}$) is a colourless silicate mineral named after the place where it was discovered, Álamos, Sonora, Mexico. It is a rare secondary mineral occurring in the oxidized zones of lead-rich deposits. For example, the infobox picture shows its association with black leadhillite.

Alarsite

Alarsite ($Al[AsO_4]$) is a mineral found in the far Eastern area of Russia. Its name is derived from its composition of Aluminium and Arsenic. Alarsite's Mohs hardness is 5-5½ and has a Trigonal crystal system.

Albite

Albite is a plagioclase feldspar mineral. It is the sodium endmember of the plagioclase solid solution series. As such it represents a plagioclase with less than 10% anorthite content. The pure albite endmember has the formula $NaAlSi_3O_8$. It is a tectosilicate. Its colour is usually pure white, hence its name from Latin *albus*.

Albite

Albite crystallizes with triclinic pinacoidal forms. Its specific gravity is about 2.62 and it has a Mohs hardness of 6 - 6.5. Albite almost always exhibits crystal twinning often as minute parallel striations on the crystal face. Albite often occurs as fine parallel segregations alternating with pink microcline in perthite as a result of exolution on cooling. It occurs in granitic and pegmatite masses, in some hydrothermal vein deposits and forms part of the typical greenschist metamorphic facies for rocks of originally basaltic composition. It was first reported in 1815 for an occurrence in Finnbo, Falun, Dalarna, Sweden. The name is from Latin, *albus* for the typical white colour.

Alforsite

Alforsite is a mineral, $Ba_5(PO_4)_3Cl$, composed of barium, phosphorus, chlorine, and oxygen. It was discovered in 1981, and named to honour geologist John T. Alfors (1930-2005) of the California Geological Survey (formerly California Division of Mines and Geology) for his work in the area where it was discovered.

Alforsite is a hexagonal colourless crystal in the chemical class phosphates and the group apatite. It is found in certain parts of central California, primarily Fresno, Mariposa, and Tulare Counties. It has also been found in Baja California, Mexico.

Algodonite

Figure 15 : Mohawkite nugget, a mixture of domeykite, algodonite and copper

Algodonite is a copper arsenide mineral with formula: Cu_6As. It is a gray white metallic mineral crystallizing in the hexagonal system. It has a Mohs hardness of 4 and a specific gravity of 8.38-8.72.

It was first described in 1857 from the Algodones silver mine, Coquimbo, Chile.

Aliettite

Aliettite is a complex phyllosilicate mineral of the smectite group with a formula of $(Ca_{0.2}Mg_6(Si,Al)_8O_{20}(OH)_4 \cdot 4H_2O)$ or $[Mg_3Si_4O_{10}(OH)_2](Ca_{0.5},Na)_{0.33}(Al,Mg,Fe^{2+})_{2-3}(Si,Al)_4O_{10}(OH)_2 \cdot n(H_2O)$.

It is a soft, colourless to pale yellow or green earthy mineral which crystallizes in the monoclinic system as minute tabular to platy crystals.

It was first described in 1968 for an occurrence in Monte Chiaro, Albareto, Parma Province, Emilia-Romagna, Italy and named for the Italian mineralogist Andrea Alietti.

It occurs in serpentinized ophiolites and their residual soil. It also occurs in altered dolostone. Associated minerals include talc, chlorite, serpentine and calcite. In addition tot the type locality in Italy it has been reported from Kinshasa, Katanga; the Chelyabinsk Oblast of the southern Urals and the Turii alkaline Massif of the Kola Peninsula in Russia; the Zirabulak Mountains of Uzbekistan; and the Goldstrike Mine of Eureka County, Nevada, USA.

Chapter 5

Crystal and Ores

Allabogdanite

Allabogdanite is a very rare phosphide mineral with formula $(Fe,Ni)_2P$, found in 1997 in a meteorite. It was described for an occurrence in the Onello meteorite in the Onello River basin, Sakha Republic; Yakutia, Russia; associated with taenite, schreibersite, kamacite, graphite and awaruite. It was named for Russian geologist Alla Bogdanova.

Allanite

Allanite is a sorosilicate group of minerals within the broader epidote group that contain a significant amount of rare earth elements. The mineral occurs mainly in metamorphosed clay rich sediments and felsic igneous rocks. It has the general formula $A_2M_3Si_3O_{12}[OH]$ where the A sites can contain large cations such as Ca^{2+}, Sr^{2+}, and rare earth elements and the M sites admit Al^{3+}, Fe^{3+}, Mn^{3+}, Fe^{2+}, or Mg^{2+} among others. However, a large amount of additional elements, including Th, U, Zr, P, Ba, Cr and others may be present in the mineral. The International Mineralogical Association lists three minerals in the allanite group, each recognized as a unique mineral: allanite-(Ce), allanite-(La) and allanite-(Y), depending on the dominant rare earth present, cerium, lanthanum or yttrium.

Allanite, also called orthite, contains up to 20% rare earth elements and is a valuable source of them. The inclusion of thorium and other radioactive elements in allanite results in some interesting phenomena. Allanite often has a pleochroic halo of radiation damage in the minerals immediately adjacent. Also highly radioactive grains of allanite often have their structure disrupted or are *metamict*. The age of allanite grains that have not been destroyed by radiation can be determined using different techniques.

Figure 1 : Allanite crystals on smokey quartz from the White Mountain Wilderness, Lincoln County, New Mexico, USA (size: 2.7×1.8×1.7 cm)

Allanite is usually black in colour, but can be brown, brown-violet. It is often coated with a yellow-brown alertation product, likely limonite. It crystallizes in the monoclinic system and forms prismatic crystals. It has a Mohs hardness of 5.5–6 and a specific gravity of 3.5–4.2. It is also pyrognomic, meaning that it becomes incandescent at a relatively low temperature of about 95 °C.

It was discovered in 1810 and named for the Scottish mineralogist, Thomas Allan (1777–1833). The type locality is Aluk Island, Greenland where it was first discovered by Karl Ludwig Giesecke.

Alloclasite

Alloclasite ($(Co,Fe)AsS$) is a sulfosalt mineral. It is a member of the arsenopyrite group. Alloclasite crystallizes in the monoclinic system and typically forms as columnar to radiating acicular prismatic clusters. It is an opaque steel-gray to silver-white, with a metallic lustre and a black streak. It is brittle with perfect cleavage, a Mohs hardness of 5 and a specific gravity of 5.91–5.95.

It was first described in 1866 for an occurrence in Romania. Its name is derived from Greek for "other" and "to break," in reference to its distinct cleavage which distinguished it from the similar appearing mineral marcasite.

Allophane

Allophane is a poorly-crystalline hydrous aluminium silicate clay *mineraloid*. Its chemical formula is $Al_2O_3\cdot(SiO_2)_{1.3-2}\cdot 2.5\text{-}3(H_2O)$. Since it has short-range atomic order, it is a mineraloid, rather than a mineral, and can be identified by its distinctive infrared spectrum and its X-ray diffraction pattern. It was first described in 1816 in Gräfenthal, Thuringia, Germany. Allophane is a weathering or hydrothermal alteration product of volcanic glass and feldspars and has a composition similar to kaolinite. It typically forms under mildly acidic to neutral pH (5-7). Its structure has been debated, but it is similar to clay minerals and is composed of curved alumina octahedral and silica tetrahedral layers. Transmission electron micrographs show that it is generally made up of aggregates of hollow spherules ~3-5 nm in diameter. Allophane can alter to form halloysite under resilicating aqueous conditions and can alter to form gibbsite under desilicating conditions. A copper containing variety *cupro-allophane* has been reported.

It forms waxy botryoidal to crusty masses with colour varying from white through green, blue, yellow, to brown. It has a Mohs hardness of 3 and a specific gravity of 1.0.

It was named from the Greek *allos*-"other" and *phanos*-"to appear", as it gave a deceptive reaction in the blowpipe flame in old mineralogical testing.

Almandine

Almandine also known incorrectly as almandite, is a species of mineral belonging to the garnet Group. The name is a corruption of alabandicus, which is the name applied by Pliny the Elder to a stone found or worked at Alabanda, a town in Caria in Asia Minor. Almandine is an iron alumina garnet, of deep red colour, inclining to purple. It is frequently cut with a convex face, or en cabochon, and is then known as carbuncle. Viewed through the spectroscope in a strong light, it generally shows three characteristic absorption bands.

Almandine is one end-member of a mineral solid solution series, with the other end member being the garnet pyrope. The almandine crystal formula is: $Fe_3Al_2(SiO_4)_3$. Magnesium substitutes for the iron with increasingly pyrope-rich composition.

Almandine, $Fe^{2+}{}_3Al_2Si_3O_{12}$, is the ferrous iron end member of the class of garnet minerals representing an important group of rock-forming silicates, which are the main constituents of the Earth's crust, upper mantle and transition zone. Almandine crystallizes in the cubic space group Ia3d, with unit-cell parameter a H" 11.512 Å at 100 K.

Almandine is antiferromagnet with the Néel temperature of 7.5 K. It contains two equivalent magnetic sublattices

Occurrence

Almandine occurs rather abundantly in the gem-gravels of Sri Lanka, whence it has sometimes been called Ceylon-ruby. When the colour inclines to a violet tint, the stone is often called Syrian garnet, a name said to be taken from Syriam, an ancient town of Pegu. Large deposits of fine almandine-garnets were found, some years ago, in the Northern Territory of Australia, and were at first taken for rubies and thus they were known in trade for some time afterwards as Australian rubies.

Almandine is widely distributed. Fine rhombic dodecahedra occur in the schistose rocks of the Zillertal, in Tyrol, and are sometimes cut and polished. An almandine in which the ferrous oxide is replaced partly by magnesia is found at Luisenfeld in German East Africa. In the United States there are many localities which yield almandine. Fine crystals of almandine embedded in mica-schist occur near Fort Wrangell in Alaska. The coarse varieties of almandine are often crushed for use as an abrasive agent.

Alstonite

Alstonite, also known as bromlite, is a member of the aragonite group of carbonate minerals. It is a calcium and barium carbonate mineral with formula, $BaCa(CO_3)_2$. It is triclinic in crystal structure and differs structurally from barytocalcite, which has a monoclinic structure. Simple crystals are not known. The crystals are invariably complex twins, and have the form of doubly terminated pseudo-hexagonal pyramids, like those of witherite but more acute; the faces are horizontally striated and are divided down their centre by a twin suture.

The examination in polarized light of a transverse section shows that each compound crystal is built up of six differently oriented individuals arranged in twelve segments.

The crystals are translucent and white, sometimes with a shade of pink. Specific gravity is 3.706, and hardness is 4-4.1.

The mineral has been found in a variety of places, two of which are in the north of England. At the Fallowfield lead mine, near

Hexham in Northumberland, it is associated with witherite; and at Bromley Hill, near Alston in Cumberland, it occurs in veins with galena. The species was named bromlite by T. Thomson in 1837, and alstonite by A. Breithaupt in 1841, both of which names, derived from the locality, have been in common use.

It occurs typically in low-temperature hydrothermal lead-zinc ore deposits and has been reported as a rare phase in carbonatites. It occurs associated with calcite, baryte, ankerite, siderite, benstonite, galena, sphalerite, pyrite and quartz.

Altaite

Altaite, or lead telluride, is a yellowish white mineral with an isometric crystal structure. Altaite is in the galena group of minerals as it shares many of properties of galena. Altaite has an unusually high density for a light-coloured mineral. Altaite and other rare tellurides are classified in the sulfide mineral class (Dana classification).

Altaite was discovered in 1845 in the Altai Mountains. Besides these mountains altaite can also be found in Zyrianovsk, Kazakhstan; the Ritchie Creek Deposit in Price County, Wisconsin; the Koch-Bulak gold deposit in Kazakhstan; Moctezuma, Mexico; and Coquimbo, Chile among other locations.

Aluminite

Aluminite is a hydrous aluminium sulfate mineral with formula: $Al_2SO_4(OH)_4 \cdot 7(H_2O)$. It is an earthy white to gray-white monoclinic mineral which almost never exhibits crystal form. It forms botryoidal to mammillary clay-like masses. It is very soft Mohs hardness of 1 with a specific gravity of 1.66—1.7. It forms as an oxidation product of pyrite and marcasite along with aluminous silicates and clays.

It was first described in 1807 from Halle, Saxony-Anhalt, Germany. It is also known as *alley stone* and *websterite* (named after English geologist Thomas Webster).

Aluminite is used by tile & masonry workers to reduce the setting time of cementitious mortars.

Aluminium

Aluminium or aluminum (US English) is a silvery white member of the boron group of chemical elements. It has the symbol Al, and its atomic number is 13. It is not soluble in water under normal circumstances. Aluminium is the third most abundant element (after oxygen and silicon), and the most abundant metal, in the Earth's crust. It makes up about 8% by weight of the Earth's solid surface. Aluminium metal is too reactive chemically to occur natively. Instead, it is found combined in over 270 different minerals. The chief ore of aluminium is bauxite.

Aluminium is remarkable for the metal's low density and for its ability to resist corrosion due to the phenomenon of passivation. Structural components made from aluminium and its alloys are vital to the aerospace industry and are important in other areas of transportation and structural materials. The most useful compounds of aluminium, at least on a weight basis, are the oxides and sulfates.

Despite its prevalence in the environment, aluminium salts are not known to be used by any form of life. In keeping with its pervasiveness, it is well tolerated by plants and animals. Because of their prevalence, potential biological roles, beneficial and otherwise, for aluminium compounds are of continuing interest.

Physical

Aluminium is a soft, durable, lightweight, ductile and malleable metal with appearance ranging from silvery to dull gray, depending on the surface roughness. Aluminium is nonmagnetic and does not easily ignite. A fresh film of aluminium film serves as a good reflector (approximately 92%) of visible light and an excellent reflector (as much as 98%) of medium and far infrared radiation. The yield strength of pure aluminium is 7–11 MPa, while aluminium alloys have yield strengths ranging from 200 MPa to 600 MPa. Aluminium has about one-third the density and stiffness of steel. It is easily machined, cast, drawn and extruded.

Corrosion resistance can be excellent due to a thin surface layer of aluminium oxide that forms when the metal is exposed to air, effectively preventing further oxidation. The strongest aluminium alloys are less corrosion resistant due to galvanic reactions with alloyed copper. This corrosion resistance is also often greatly reduced when many aqueous salts are present, particularly in the presence of dissimilar metals.

Aluminium atoms are arranged in a face-centred cubic (fcc) structure. Aluminium has a stacking-fault energy of approximately 200 mJ/m^2.

Aluminium is one of the few metals that retain full silvery reflectance in finely powdered form, making it an important component of silver-coloured paints. Aluminium mirror finish has the highest reflectance of any metal in the 200–400 nm (UV) and the 3,000–10,000 nm (far IR) regions; in the 400–700 nm visible range it is slightly outperformed by tin and silver and in the 700–3000 (near IR) by silver, gold, and copper.

Aluminium is a good thermal and electrical conductor, having 59% the conductivity of copper, both thermal and electrical. Aluminium is capable of being a superconductor, with a superconducting critical temperature of 1.2 Kelvin and a critical magnetic field of about 100 gauss (10 milliteslas).

Creation

Stable aluminium is created when hydrogen fuses with magnesium either in large stars or in supernovae.

Isotopes

Aluminium has many known isotopes, whose mass numbers range from 21 to 42; however, only ^{27}Al (stable isotope) and ^{26}Al (radioactive isotope, $t_{1/2} = 7.2 \times 10^5$ y) occur naturally. ^{27}Al has a natural abundance above 99.9%. ^{26}Al is produced from argon in the atmosphere by spallation caused by cosmic-ray protons. Aluminium isotopes have found practical application in dating marine sediments, manganese nodules, glacial ice, quartz in rock exposures, and meteorites. The ratio of ^{26}Al to ^{10}Be has been used to study the role of transport, deposition, sediment storage, burial times, and erosion on 10^5 to 10^6 year time scales. Cosmogenic ^{26}Al was first applied in studies of the Moon and meteorites. Meteoroid fragments, after departure from their parent bodies, are exposed to intense cosmic-ray bombardment during their travel through space, causing substantial ^{26}Al production. After falling to Earth, atmospheric shielding drastically reduces ^{26}Al production, and its decay can then be used to determine the meteorite's terrestrial age. Meteorite research has also shown that ^{26}Al was relatively abundant at the time of formation of our planetary system. Most meteorite scientists believe that the energy released by the decay of ^{26}Al was responsible for the melting and differentiation of some asteroids after their formation 4.55 billion years ago.

Natural Occurrence

In the Earth's crust, aluminium is the most abundant (8.3% by weight) metallic element and the third most abundant of all elements (after oxygen and silicon). Because of its strong affinity to oxygen, it is almost never found in the elemental state; instead it is found in oxides or silicates. Feldspars, the most common group of minerals in the Earth's crust, are aluminosilicates. Native aluminium metal can be found as a minor phase in low oxygen fugacity environments, such as the interiors of certain volcanoes. Native aluminium has been reported in cold seeps in the northeastern continental slope of the South China Sea and Chen et al. (2011) have proposed a theory of its origin as resulting by reduction from tetrahydroxoaluminate $Al(OH)_4^-$ to metallic aluminium by bacteria. It also occurs in the minerals beryl, cryolite, garnet, spinel and turquoise. Impurities in Al_2O_3, such as chromium or iron yield the gemstones ruby and sapphire, respectively. Although aluminium is an extremely common and widespread element, the common aluminium minerals are not economic sources of the metal. Almost all metallic aluminium is produced from the ore bauxite ($AlO_x(OH)_{3-2x}$). Bauxite occurs as a weathering product of low iron and silica bedrock in tropical climatic conditions. Large deposits of bauxite occur in Australia, Brazil, Guinea and Jamaica and the primary mining areas for the ore are in Australia, Brazil, China, India, Guinea, Indonesia, Jamaica, Russia and Surinam.

Production and Refinement

Although aluminium is the most abundant metallic element in the Earth's crust, it is never found in free, metallic form.

Figure 2 : Bauxite, a major aluminium ore. The red-brown colour is due to the presence of iron minerals.

Aluminium forms strong chemical bonds with oxygen. Compared to most other metals, it is difficult to extract from ore, such as bauxite, due to the energy required to reduce aluminium oxide (Al_2O_3). For example, direct reduction with carbon, as is used to produce iron, is not chemically possible, since aluminium is a stronger reducing agent than carbon. There is an indirect carbothermic reduction possible by using carbon and Al_2O_3, which forms an intermediate Al_4C_3 and this can further yield aluminium metal at a temperature of 1900–2000°C. This process is still under development. This process costs less energy and yields less CO_2 than the Hall-Héroult process, the major industrial process for aluminium extraction. Aluminium oxide has a melting point of about 2,000 °C (3,600 °F). Therefore, it must be extracted by electrolysis. In this process, the aluminium oxide is dissolved in molten cryolite with calcium fluoride and then reduced to the pure metal. The operational temperature of the reduction cells is around 950 to 980 °C (1,740 to 1,800 °F). Cryolite is found as a mineral in Greenland, but in industrial use it has been replaced by a synthetic substance. Cryolite is a chemical compound of aluminium and sodium fluorides: (Na_3AlF_6). The aluminium oxide (a white powder) is obtained by refining bauxite in the Bayer process of Karl Bayer. (Previously, the Deville process was the predominant refining technology.)

The electrolytic process replaced the Wöhler process, which involved the reduction of anhydrous aluminium chloride with potassium. Both of the electrodes used in the electrolysis of aluminium oxide are carbon. Once the refined alumina is dissolved in the electrolyte, its ions are free to move around. The reaction at the cathode is:

$$Al^{3+} + 3\ e^- \rightarrow Al$$

Here the aluminium ion is being reduced. The aluminium metal then sinks to the bottom and is tapped off, usually cast into large blocks called aluminium billets for further processing.

At the anode, oxygen is formed:

$$2\ O^{2-} \rightarrow O_2 + 4\ e^-$$

To some extent, the carbon anode is degraded by the oxygen. The anodes in a reduction cell must therefore be replaced regularly, since they are consumed in the process. The cathodes do erode, mainly due to electrochemical processes and metal movement. After five to ten years, depending on the current used in the electrolysis, a cell has to be rebuilt because of cathode wear.

Aluminium electrolysis with the Hall-Héroult process consumes a lot of energy, but alternative processes were always found to be less viable economically and/or ecologically.

The worldwide average specific energy consumption is approximately 15±0.5 kilowatt-hours per kilogram of aluminium produced (52 to 56 MJ/kg).

The most modern smelters achieve approximately 12.8 kW·h/kg (46.1 MJ/kg). (Compare this to the heat of reaction, 31 MJ/kg, and the Gibbs free energy of reaction, 29 MJ/kg.) Reduction line currents for older technologies are typically 100 to 200 kiloamperes; state-of-the-art smelters operate at about 350 kA. Trials have been reported with 500 kA cells.

The Hall-Heroult process produces aluminium with a purity of above 99%. Further purification can be done by the Hoope process. The process involves the electrolysis of molten aluminium with a sodium, barium and aluminium fluoride electrolyte. The resulting aluminium has a purity of 99.99%.

Electric power represents about 20% to 40% of the cost of producing aluminium, depending on the location of the smelter.

Aluminium production consumes roughly 5% of electricity generated in the U.S. Smelters tend to be situated where electric power is both plentiful and inexpensive, such as the United Arab Emirates with excess natural gas supplies and Iceland and Norway with energy generated from renewable sources. The world's largest smelters of alumina are People's Republic of China, Russia, and Quebec and British Columbia in Canada.

In 2005, the People's Republic of China was the top producer of aluminium with almost a one-fifth world share, followed by Russia, Canada, and the USA, reports the British Geological Survey.

Over the last 50 years, Australia has become a major producer of bauxite ore and a major producer and exporter of alumina (before being overtaken by China in 2007). Australia produced 62 million tonnes of bauxite in 2005. The Australian deposits have some refining problems, some being high in silica, but have the advantage of being shallow and relatively easy to mine.

Recycling

Aluminium is theoretically 100% recyclable without any loss of its natural qualities. According to the International Resource Panel's

Metal Stocks in Society report, the global per capita stock of aluminium in use in society (i.e. in cars, buildings, electronics etc.) is 80 kg. Much of this is in more-developed countries (350–500 kg per capita) rather than less-developed countries (35 kg per capita). Knowing the per capita stocks and their approximate lifespans is important for planning recycling.

Figure 3 : Aluminium Recycling code

Recovery of the metal via recycling has become an important facet of the aluminium industry. Recycling was a low-profile activity until the late 1960s, when the growing use of aluminium beverage cans brought it to the public awareness.

Recycling involves melting the scrap, a process that requires only 5% of the energy used to produce aluminium from ore, though a significant part (up to 15% of the input material) is lost as dross (ash-like oxide). The dross can undergo a further process to extract aluminium.

In Europe aluminium experiences high rates of recycling, ranging from 42% of beverage cans, 85% of construction materials and 95% of transport vehicles.

Recycled aluminium is known as secondary aluminium, but maintains the same physical properties as primary aluminium. Secondary aluminium is produced in a wide range of formats and is employed in 80% of alloy injections. Another important use is for extrusion.

White dross from primary aluminium production and from secondary recycling operations still contains useful quantities of aluminium that can be extracted industrially. The process produces aluminium billets, together with a highly complex waste material. This waste is difficult to manage. It reacts with water, releasing a mixture of gases (including, among others, hydrogen, acetylene, and ammonia), which spontaneously ignites on contact with air; contact with damp air results in the release of copious quantities of ammonia gas. Despite these difficulties, the waste has found use as a filler in asphalt and concrete.

Compounds

Oxidation State +3

The vast majority of compounds, including all Al-containing minerals and all commercially significant aluminium compounds, feature aluminium in the oxidation state 3+. The coordination number of such compounds varies, but generally Al^{3+} is six-coordinate or tetracoordinate. Almost all compounds of aluminium(III) are colourless.

Halides

All four trihalides are well known. Unlike the structures of the three heavier trihalides, aluminium fluoride (AlF_3) features six-coordinate Al. The octahedral coordination environment for AlF_3 is related to the compactness of fluoride ion, six of which can fit around the small Al^{3+} centre. AlF_3 sublimes (with cracking) at 1,291 °C (2,356 °F). With heavier halides, the coordination numbers are lower. The other trihalides are dimeric or polymeric with tetrahedral Al centres.

These materials are prepared by treating aluminium metal with the halogen, although other methods exist. Acidification of the oxides or hydroxides affords hydrates. In aqueous solution, the halides often form mixtures, generally containing six-coordinate Al centres, which are feature both halide and aquo ligands. When aluminium and fluoride are together in aqueous solution, they readily form complex ions such as $[AlF(H_2O)_5]^{2+}$, $AlF_3(H_2O)_3$, and $[AlF_6]^{3-}$. In the case of chloride, polyaluminium clusters are formed such as $[Al_{13}O_4(OH)_{24}(H_2O)_{12}]^{7+}$.

Oxide and Hydroxides

Aluminium forms one stable oxide, known by its mineral name corundum. Sapphire and ruby are impure corundum contaminated with trace amounts of other metals. The two oxide-hydroxides (AlO(OH) are boehmite and diaspore.

There are three trihydroxides: bayerite, gibbsite, and nordstrandite, which differ in their crystalline structure (polymorphs). Most are produced from ores by a variety of wet processes using acid and base. Heating the hydroxides leads to formation of corundrum. These materials are of central importance to the production of aluminium and are themselves extremely useful.

Carbide, Nitride, and Related Materials

Aluminium carbide (Al_4C_3) is made by heating a mixture of the elements above 1,000 °C (1,832 °F). The pale yellow crystals consist of tetrahedral aluminium centres. It reacts with water or dilute acids to give methane. The acetylide, $Al_2(C_2)_3$, is made by passing acetylene over heated aluminium.

Aluminium nitride (AlN) is the only nitride known for aluminium. Unlike the oxides it features tetrahedral Al centres. It can be made from the elements at 800 °C (1,472 °F). It is air-stable material with a usefully high thermal conductivity. Aluminium phosphide (AlP) is made similarly, and hydrolyses to give phosphine:

$$AlP + 3\ H_2O \rightarrow Al(OH)_3 + PH_3$$

Organoaluminium Compounds and Related Hydrides

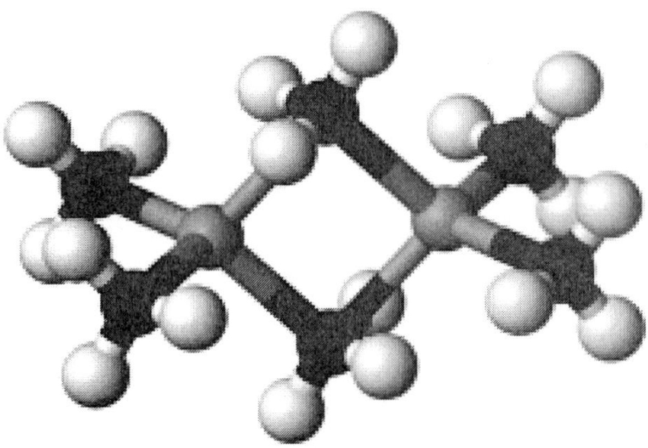

Figure 4 : Structure of trimethylaluminium, a compound that features five-coordinate carbon.

A variety of compounds of empirical formula AlR_3 and $AlR_{1.5}Cl_{1.5}$ exist. These species usually feature tetrahedral Al centres, e.g. "trimethylaluminium" has the formula $Al_2(CH_3)_6$. With large organic groups, triorganoaluminium exist as three-coordinate monomers, such as triisobutylaluminium. Such compounds are widely used in industrial chemistry, despite the fact that they are often highly pyrophoric. Few analogues exist between organoaluminium and organoboron compounds except for large organic groups.

The important aluminium hydride is lithium aluminium hydride ($LiAlH_4$), which is used in as a reducing agent in organic chemistry. It can be produced from lithium hydride and aluminium trichloride:

$$4\ LiH + AlCl_3 \rightarrow LiAlH_4 + 3\ LiCl$$

Several useful derivatives of $LiAlH_4$ are known, e.g. Sodium bis(2-methoxyethoxy)dihydridoaluminate. The simplest hydride, aluminium hydride or alane, remains a laboratory curiosity. It is a polymer with the formula $(AlH_3)_n$, in contrast to the corresponding boron hydride with the formula $(BH_3)_2$.

Oxidation States +1 and +2

Although the great majority of aluminium compounds feature Al^{3+} centres, compounds with lower oxidation states are known and sometime of significance as precursors to the Al^{3+} species.

Aluminium (I)

AlF, AlCl and AlBr exist in the gaseous phase when the tri-halide is heated with aluminium. The composition AlI is unstable at room temperature with respect to the triiodide:

$$3\ AlI \rightarrow AlI_3 + 2\ Al$$

A stable derivative of aluminium monoiodide is the cyclic adduct formed with triethylamine, $Al_4I_4(NEt_3)_4$. Also of theoretical interest but only of fleeting existence are Al_2O and Al_2S is made by heating the normal oxide, Al_2O_3, with silicon at 1,800 °C (3,272 °F) in a vacuum. Such materials quickly disproportionates to the starting materials.

Aluminium (II)

Very simple Al(II) compounds are invoked or observed in the reactions of Al metal with oxidants. For example, aluminium monoxide, AlO, has been detected in the gas phase after explosion and in stellar absorption spectra. More thoroughly investigated are compounds of the formula R_4Al_2 where R is a large organic ligand.

Analysis

The presence of aluminium can be detected in qualitative analysis using aluminon.

Applications

General use

Aluminium is the most widely used non-ferrous metal. Global production of aluminium in 2005 was 31.9 million tonnes. It exceeded that of any other metal except iron (837.5 million tonnes). Forecast for 2012 is 42–45 million tons, driven by rising Chinese output.

Aluminium is almost always alloyed, which markedly improves its mechanical properties, especially when tempered. For example, the common aluminium foils and beverage cans are alloys of 92% to 99% aluminium.

The main alloying agents are copper, zinc, magnesium, manganese, and silicon (e.g., duralumin) and the levels of these other metals are in the range of a few percent by weight.

Some of the many uses for aluminium metal are in:

- Transportation (automobiles, aircraft, trucks, railway cars, marine vessels, bicycles, etc.) as sheet, tube, castings, etc.
- Packaging (cans, foil, etc.)
- Construction (windows, doors, siding, building wire, etc.)
- A wide range of household items, from cooking utensils to baseball bats, watches.
- Street lighting poles, sailing ship masts, walking poles, etc.
- Outer shells of consumer electronics, also cases for equipment e.g. photographic equipment.
- Electrical transmission lines for power distribution
- MKM steel and Alnico magnets
- Super purity aluminium (SPA, 99.980% to 99.999% Al), used in electronics and CDs.
- Heat sinks for electronic appliances such as transistors and CPUs.
- Substrate material of metal-core copper clad laminates used in high brightness LED lighting.
- Powdered aluminium is used in paint, and in pyrotechnics such as solid rocket fuels and thermite.
- Aluminium can be reacted with hydrochloric acid or with sodium hydroxide to produce hydrogen gas.
- A variety of countries, including France, Italy, Poland, Finland, Romania, Israel, and the former Yugoslavia, have issued coins struck in aluminium or aluminium-copper alloys.
- Some guitar models sports aluminium diamond plates on the surface of the instruments, usually either chrome or black. Kramer Guitars and Travis Bean are both known for having produced guitars with necks made of aluminium, which gives the instrument a very distinct sound.

- Sustainability of Aluminium in Buildings.

Relatively pure aluminium is rarely encountered – only when corrosion resistance and/or workability is more important than strength or hardness. A thin layer of aluminium can be deposited onto a flat surface by physical vapour deposition or (very infrequently) chemical vapour deposition or other chemical means to form optical coatings and mirrors.

Aluminium Compounds

Because aluminium is abundant and most of its derivative exhibit low toxicity, the compounds of aluminium enjoy wide and sometimes large-scale applications.

Alumina

Aluminium oxide (Al_2O_3) and the associated oxy-hydroxides and trihydroxides are produced or extracted from minerals on a large scale. The great majority of this material is converted to metallic aluminium. About 10% of the production capacity is used for other applications. A major use is as an absorbent, for example alumina will remove water from hydrocarbons, to enable subsequent processes that are poisoned by moisture. Aluminium oxides are common catalysts for industrial processes, e.g. the Claus process for converting hydrogen sulfide to sulfur in refineries and for the alkylation of amines. Many industrial catalysts are "supported", meaning generally that an expensive catalyst (e.g., platinum) is dispersed over a high surface area material such as alumina. Being a very hard material (Mohs hardness 9), alumina is widely used as an abrasive and the production of applications that exploit its inertness, e.g., in high pressure sodium lamps.

Sulfates

Several sulfates of aluminium find applications. Aluminium sulfate ($Al_2(SO_4)_3(H_2O)_{18}$) is produced on the annual scale of several billions of kilograms. About half of the production is consumed in water treatment. The next major application is in the manufacture of paper. It is also used as a mordant, in fire extinguisher, as a food additive, in fireproofing, and in leather tanning. Aluminium ammonium sulfate ($[Al(NH_4)](SO_4)_2)(H_2O)_{12}$, also called ammonium alum, is used as a mordant and in leather tanning. Aluminium potassium sulfate ($[Al(K)](SO_4)_2)(H_2O)_{12}$ is used similarly. The consumption of both alums is declining.

Chlorides

Aluminium chloride ($AlCl_3$) is used in petroleum refining and in the production of synthetic rubber and polymers. Although it has a similar name, aluminium chlorohydrate has fewer and very different applications, e.g. as an hardening agent and an antiperspirant. It is an intermediate in the production of aluminium metal.

Niche Compounds

Given the scale of aluminium compounds, a small scale application could still involve thousands of tons. One of the many compounds used at this intermediate level include aluminium acetate, a salt used in solution as an astringent. Aluminium borate (Al_2O_3 B_2O_3) is used in the production of glass and ceramics. Aluminium fluorosilicate ($Al_2(SiF_6)_3$) is used in the production of synthetic gemstones, glass and ceramic. Aluminium phosphate ($AlPO_4$) is used in the manufacture: of glass and ceramic, pulp and paper products, cosmetics, paints and varnishes and in making dental cement. Aluminium hydroxide ($Al(OH)_3$) is used as an antacid, as a mordant, in water purification, in the manufacture of glass and ceramic and in the waterproofing of fabrics. Lithium aluminium hydride is a powerful reducing agent used in organic chemistry. Organoaluminiums are used as Lewis acids and cocatalysts. For example, methylaluminoxane is a cocatalyst for Ziegler-Natta olefin polymerization to produce vinyl polymers such as polyethene.

Aluminium Alloys in Structural Applications

Aluminium alloys with a wide range of properties are used in engineering structures. Alloy systems are classified by a number system (ANSI) or by names indicating their main alloying constituents (DIN and ISO).

The strength and durability of aluminium alloys vary widely, not only as a result of the components of the specific alloy, but also as a result of heat treatments and manufacturing processes. A lack of knowledge of these aspects has from time to time led to improperly designed structures and gained aluminium a bad reputation.

One important structural limitation of aluminium alloys is their fatigue strength. Unlike steels, aluminium alloys have no well-defined fatigue limit, meaning that fatigue failure eventually occurs, under even very small cyclic loadings. This implies that engineers must assess these loads and design for a fixed life rather than an infinite life.

Another important property of aluminium alloys is their sensitivity to heat. Workshop procedures involving heating are complicated by the fact that aluminium, unlike steel, melts without first glowing red. Forming operations where a blow torch is used therefore require some expertise, since no visual signs reveal how close the material is to melting. Aluminium alloys, like all structural alloys, also are subject to internal stresses following heating operations such as welding and casting.

The problem with aluminium alloys in this regard is their low melting point, which make them more susceptible to distortions from thermally induced stress relief. Controlled stress relief can be done during manufacturing by heat-treating the parts in an oven, followed by gradual cooling—in effect annealing the stresses.

The low melting point of aluminium alloys has not precluded their use in rocketry; even for use in constructing combustion chambers where gases can reach 3500 K. The Agena upper stage engine used a regeneratively cooled aluminium design for some parts of the nozzle, including the thermally critical throat region.

Another alloy of some value is aluminium bronze (CuAl5).

History

Ancient Greeks and Romans used aluminium salts as dyeing mordants and as astringents for dressing wounds; alum is still used as a styptic. In 1761, Guyton de Morveau suggested calling the base alum *alumine.* In 1808, Humphry Davy identified the existence of a metal base of alum, which he at first termed *alumium* and later *aluminum*.

The metal was first produced in 1825 (in an impure form) by Danish physicist and chemist Hans Christian Orsted. He reacted anhydrous aluminium chloride with potassium amalgam and yielded a lump of metal looking similar to tin.

Friedrich Wöhler was aware of these experiments and cited them, but after redoing the experiments of Orsted he concluded that this metal was pure potassium. He conducted a similar experiment in 1827 by mixing anhydrous aluminium chloride with potassium and yielded aluminium. Wöhler is generally credited with isolating aluminium (Latin *alumen*, alum), but also Orsted can be listed as its discoverer. Further, Pierre Berthier discovered aluminium in bauxite ore and successfully extracted it. Frenchman Henri Etienne Sainte-Claire

Deville improved Wöhler's method in 1846, and described his improvements in a book in 1859, chief among these being the substitution of sodium for the considerably more expensive potassium. Deville likely also conceived the idea of the electrolysis of aluminium oxide dissolved in cryolite; Charles Martin Hall and Paul Héroult might have developed the more practical process after Deville.

Before the Hall-Héroult process was developed in the late 1880s, aluminium was exceedingly difficult to extract from its various ores. This made pure aluminium more valuable than gold. Bars of aluminium were exhibited at the Exposition Universelle of 1855. Napoleon III, Emperor of France, is reputed to have given a banquet where the most honoured guests were given aluminium utensils, while the others made do with gold.

Aluminium was selected as the material to be used for the 100 ounce (2.8 kg) capstone of the Washington Monument in 1884, a time when one ounce (30 grams) cost the daily wage of a common worker on the project; The capstone, which was set in place on December 6, 1884, in an elaborate dedication ceremony, was the largest single piece of aluminium cast at the time, when aluminium was as expensive as silver.

The Cowles companies supplied aluminium alloy in quantity in the United States and England using smelters like the furnace of Carl Wilhelm Siemens by 1886. Charles Martin Hall of Ohio in the U.S. and Paul Héroult of France independently developed the Hall-Héroult electrolytic process that made extracting aluminium from minerals cheaper and is now the principal method used worldwide. Hall's process, in 1888 with the financial backing of Alfred E. Hunt, started the Pittsburgh Reduction Company today known as Alcoa. Héroult's process was in production by 1889 in Switzerland at Aluminium Industrie, now Alcan, and at British Aluminium, now Luxfer Group and Alcoa, by 1896 in Scotland.

By 1895, the metal was being used as a building material as far away as Sydney, Australia in the dome of the Chief Secretary's Building.

Many navies have used an aluminium superstructure for their vessels; the 1975 fire aboard USS *Belknap* that gutted her aluminium superstructure, as well as observation of battle damage to British ships during the Falklands War, led to many navies switching to all steel superstructures. The *Arleigh Burke* class was the first such U.S. ship, being constructed entirely of steel.

Aluminium wire was once widely used for domestic electrical wiring. Owing to corrosion-induced failures, a number of fires resulted. This discontinuation thus illustrates one failed application of the otherwise highly useful metal.

In 2008, the price of aluminium peaked at $1.45/lb in July but dropped to $0.70/lb by December.

Etymology

Two variants of the metal's name are in current use, *aluminium* and *aluminum* (besides the obsolete *alumium*). The International Union of Pure and Applied Chemistry (IUPAC) adopted *aluminium* as the standard international name for the element in 1990 but, three years later, recognized *aluminum* as an acceptable variant.

Hence their periodic table includes both. IUPAC prefers the use of *aluminium* in its internal publications, although nearly as many IUPAC publications use the spelling *aluminum*.

Most countries use the spelling *aluminium*. In the United States, the spelling *aluminum* predominates. The Canadian Oxford Dictionary prefers *aluminum*, whereas the Australian Macquarie Dictionary prefers *aluminium*. In 1926, the American Chemical Society officially decided to use *aluminum* in its publications; American dictionaries typically label the spelling *aluminium* as a British variant.

The name *aluminium* derives from its status as a base of alum. It is borrowed from Old French; its ultimate source, *alumen*, in turn is a Latin word that literally means "bitter salt".

The earliest citation given in the Oxford English Dictionary for any word used as a name for this element is *alumium*, which British chemist and inventor Humphry Davy employed in 1808 for the metal he was trying to isolate electrolytically from the mineral *alumina*.

The citation is from the journal *Philosophical Transactions of the Royal Society of London*: "Had I been so fortunate as to have obtained more certain evidences on this subject, and to have procured the metallic substances I was in search of, I should have proposed for them the names of silicium, alumium, zirconium, and glucium."

Davy settled on *aluminum* by the time he published his 1812 book *Chemical Philosophy*: "This substance appears to contain a peculiar metal, but as yet Aluminum has not been obtained in a perfectly free state, though alloys of it with other metalline substances have been procured sufficiently distinct to indicate the probable nature of alumina." But the same year, an anonymous contributor to the *Quarterly Review,* a British political-literary journal, in a review of Davy's book, objected to *aluminum* and proposed the name *aluminium,* "for so we shall take the liberty of writing the word, in preference to aluminum, which has a less classical sound."

The-*ium* suffix conformed to the precedent set in other newly discovered elements of the time: potassium, sodium, magnesium, calcium, and strontium (all of which Davy isolated himself). Nevertheless,-*um* spellings for elements were not unknown at the time, as for example platinum, known to Europeans since the 16th century, molybdenum, discovered in 1778, and tantalum, discovered in 1802. The-*um* suffix is consistent with the universal spelling alumina for the oxide, as lanthana is the oxide of lanthanum, and magnesia, ceria, and thoria are the oxides of magnesium, cerium, and thorium respectively.

The spelling used throughout the 19th century by most U.S. chemists was *aluminium*, but common usage is less clear. The *aluminum* spelling is used in the Webster's Dictionary of 1828. In his advertising handbill for his new electrolytic method of producing the metal 1892, Charles Martin Hall used the-*um* spelling, despite his constant use of the-*ium* spelling in all the patents he filed between 1886 and 1903. It has consequently been suggested that the spelling reflects an easier to pronounce word with one fewer syllable, or that the spelling on the flier was a mistake. Hall's domination of production of the metal ensured that the spelling *aluminum* became the standard in North America; the *Webster Unabridged Dictionary* of 1913, though, continued to use the-*ium* version.

Health Concerns

Despite its natural abundance, aluminium has no known function in biology. It is remarkably nontoxic, aluminium sulfate having an LD50 of 6207 mg/kg (oral, mouse), which corresponds to 500 grams for a 80 kg person. Despite the extremely low acute toxicity, the health effects of aluminium are of interest in view of the widespread occurrence of the element in the environment and in commerce.

Some toxicity can be traced to deposition in bone and the central nervous system, which is particularly increased in patients with reduced renal function. Because aluminium competes with calcium for absorption, increased amounts of dietary aluminium may contribute to the reduced skeletal mineralization (osteopenia) observed in preterm infants and infants with growth retardation. In very high doses, aluminium can cause neurotoxicity, and is associated with altered function of the blood-brain barrier.

A small percentage of people are allergic to aluminium and experience contact dermatitis, digestive disorders, vomiting or other symptoms upon contact or ingestion of products containing aluminium, such as deodorants or antacids.

In those without allergies, aluminium is not as toxic as heavy metals, but there is evidence of some toxicity if it is consumed in excessive amounts. Although the use of aluminium cookware has not been shown to lead to aluminium toxicity in general, excessive consumption of antacids containing aluminium compounds and excessive use of aluminium-containing antiperspirants provide more significant exposure levels. Studies have shown that consumption of acidic foods or liquids with aluminium significantly increases aluminium absorption, and maltol has been shown to increase the accumulation of aluminium in nervous and osseus tissue. Furthermore, aluminium increases estrogen-related gene expression in human breast cancer cells cultured in the laboratory. The estrogen-like effects of these salts have led to their classification as a metalloestrogen.

The effects of aluminium in antiperspirants has been examined over the course of decades with little evidence of skin irritation. Nonetheless, its occurrence in antiperspirants, dyes (such as aluminium lake), and food additives is controversial in some quarters. Although there is little evidence that normal exposure to aluminium presents a risk to healthy adults, some studies point to risks associated with increased exposure to the metal. Aluminium in food may be absorbed more than aluminium from water. Some researchers have expressed concerns that the aluminium in antiperspirants may increase the risk of breast cancer, and aluminium has controversially been implicated as a factor in Alzheimer's disease. The Camelford water pollution incident involved a number of people consuming aluminium sulfate. Investigations of the long-term health effects are still ongoing, but elevated brain aluminium concentrations have been found in post-mortem examinations of victims, and further research to determine if there is a link with cerebral amyloid angiopathy has been commissioned.

According to The Alzheimer's Society, the overwhelming medical and scientific opinion is that studies have not convincingly demonstrated a causal relationship between aluminium and Alzheimer's disease. Nevertheless, some studies, such as those on the PAQUID cohort, cite aluminium exposure as a risk factor for Alzheimer's disease. Some brain plaques have been found to contain increased levels of the metal. Research in this area has been inconclusive; aluminium accumulation may be a consequence of the disease rather than a causal agent. In any event, if there is any toxicity of aluminium, it must be via a very specific mechanism, since total human exposure to the element in the form of naturally occurring clay in soil and dust is enormously large over a lifetime. Scientific consensus does not yet exist about whether aluminium exposure could directly increase the risk of Alzheimer's disease.

Effect on Plants

Aluminium is primary among the factors that reduce plant growth on acid soils. Although it is generally harmless to plant growth in pH-neutral soils, the concentration in acid soils of toxic Al^{3+} cations increases and disturbs root growth and function.

Most acid soils are saturated with aluminium rather than hydrogen ions. The acidity of the soil is therefore a result of hydrolysis of aluminium compounds. This concept of "corrected lime potential" to define the degree of base saturation in soils became the basis for procedures now used in soil testing laboratories to determine the "lime requirement" of soils.

Wheat's adaptation to allow aluminium tolerance is such that the aluminium induces a release of organic compounds that bind to the harmful aluminium cations. Sorghum is believed to have the same tolerance mechanism. The first gene for aluminium tolerance has been identified in wheat. It was shown that sorghum's aluminium tolerance is controlled by a single gene, as for wheat. This is not the case in all plants.

Alunite

Alunite is a sulfate mineral that was first observed in the 15th century in Monti della Tolfa, north of Rome, where it was mined for the manufacture of alum. First called *aluminilite* by J.C. Delamétherie in 1797, this name was contracted by François Beudant in 1824 to alunite.

Distinct crystals of alunite are rarely found in cavities in the massive material. Alunite crystallizes in the hexagonal system with crystals forming trigonal pyramids that resemble rhombohedra with interfacial angles of 90° 50', so that they resemble cubes in appearance. Minute glistening crystals have also been found loose in cavities in altered rhyolite. Alunite varies in colour from white to yellow gray. The hardness is 4 and the specific gravity is between 2.6 and 2.8. The mineral is a hydrated aluminium potassium sulfate, $KAl_3(SO_4)_2(OH)_6$. Sodium substitutes for potassium and when high in sodium it is called natroalunite. It is insoluble in water or weak acids, but soluble in sulfuric acid. Jarosite is an iron analogue in which Fe^{3+} replaces the aluminium. It occurs as a secondary mineral on iron sulfate ores.

Figure 5 : Alunite from Slovakia

Figure 6 : Alunite Marysvale, Utah

Alunite occurs as veins and replacement masses in trachyte, rhyolite, and similar potassium rich volcanic rocks. It is formed by the action of sulfuric acid bearing solutions on these rocks during the oxidation and leaching of metal sulfide deposits. Alunite also is found near volcanic fumaroles. The white, finely granular masses closely resemble finely granular limestone, dolomite, anhydrite, and magnesite in appearance. The more compact kinds from Hungary are so hard and tough that they have been used for millstones.

Historically extensive deposits were mined in Tuscany and Hungary, and at Bulahdelah, Australia. Currently found at Tolfa, Italy. In the United States it is found in the San Juan district of Colorado; Goldfield, Nevada; the ghost town of Alunite, Utah near Marysvale; and Red Mountain near Patagonia, Arizona. The Arizona occurrence lies appropriately above a canyon named *Alum Gulch*. Alunite is mined as an ore of both potassium and aluminium at Marysvale. Some of the ore deposits were located by airborne and satellite multispectral imaging.

Alunogen

Alunogen also called feather alum and hair salt is a colourless to white (although often coloured by impurities, like iron substituting for aluminum) fibrous to needle-like aluminium sulfate mineral often found on the walls of mines and quarries as secondary mineral. It can be found in the oxidation zones of some ore deposits as well as on burning coal dumps (i.e. as the product of millosevichite hydratation).

Alunogen has the chemical formula $Al_2(SO_4)_3 \cdot 17H_2O$. More precise, crystallochemical formula, can be written as: $[Al(H_2O)_6]_2(SO_4)_3 \cdot 5H_2O$. The second formula shows that H_2O in the alunogen formula occurs both as ligand (coordinative form) and loosely-bounded (crystallization) form. The second form can be easily removed just by slight grinding of the mineral (by preserving the unique structure).

Amblygonite

Amblygonite is a fluorophosphate mineral, $(Li,Na)AlPO_4(F,OH)$, composed of lithium, sodium, aluminium, phosphate, fluoride and hydroxide. The mineral occurs in pegmatite deposits and is easily mistaken for albite and other feldspars. Its density, cleavage and flame test for lithium are diagnostic. Amblygonite forms a series with *montebrasite*, the low fluorine endmember. Geologic occurrence is in granite pegmatites, high-temperature tin veins, and greisens. Amblygonite occurs with spodumene, apatite, lepidolite, tourmaline, and other lithium-bearing minerals in pegmatite veins. It contains about 10% lithium, and has been utilized as a source of lithium. The chief commercial sources have historically been the deposits of California and France.

History

The mineral was first discovered in Saxony by August Breithaupt in 1817, and named by him from the Greek *amblus*, blunt, and *gouia*, angle, because of the obtuse angle between the cleavages. Later it was found at Montebras, Creuse, France, and at Hebron in Maine; and because of slight differences in optical character and chemical composition the names montebrasite and hebronite have been applied to the mineral from these localities. It has been discovered in considerable quantity at Pala in San Diego county, California; Caceres, Spain; and the Black Hills of South Dakota. The largest documented single crystal of amblygonite measured 7.62x2.44x1.83 m^3 and weighed ~102 tons.

Gemology

Transparent amblygonite has been faceted and used as a gemstone. As a gemstone set into jewellery it is vulnerable to breakage and abrasion from general wear, as its hardness and toughness are poor. The main sources for gem material are Brazil and the U.S.. Australia, France, Germany, Namibia, Norway, and Spain have also produced gem quality amblygonite.

Ameghinite

Ameghinite ($Na[H_4B_3O_7]$ or $NaB_3O_3(OH)_4$) is a mineral found in Argentina. It is a soft mineral with a Mohs hardness of 2-3. Ameghinite has a Monoclinic crystal system.

It was first described in 1967 for an occurrence in the Tincalayu Mine, Salar Del Hombre Muerto, Salta, Argentina. It was named for Argentine geologist brothers, Carlos Ameghino (1865–1936) and Florentino Ameghino (1854–1911).

Amphibole

Amphibole is the name of an important group of generally dark-coloured rock-forming inosilicate minerals, composed of double chain SiO_4 tetrahedra, linked at the vertices and generally containing ions of iron and/or magnesium in their structures.

Mineralogy

Amphiboles crystallize into two crystal systems, monoclinic and orthorhombic. In chemical composition and general characteristics they are similar to the pyroxenes. The chief differences from pyroxenes are that (i) amphiboles contain essential hydroxyl (OH) or halogen (F, Cl) and (ii) the basic structure is a double chain of tetrahedra (as opposed to the single chain structure of pyroxene). Most apparent, in hand specimens, is that amphiboles form oblique cleavage planes (at around 120 degrees), whereas pyroxenes have cleavage angles of approximately 90 degrees. Amphiboles are also specifically less dense than the corresponding pyroxenes. In optical characteristics, many amphiboles are distinguished by their stronger pleochroism and by the smaller angle of extinction (Z angle c) on the plane of symmetry. Amphiboles are the primary constituent of amphibolites.

In rocks

Amphiboles are minerals of either igneous or metamorphic origin; in the former case occurring as constituents (hornblende) of igneous rocks, such as granite, diorite, andesite and others. Calcium is sometimes a constituent of naturally occurring amphiboles.(C. Michael Hogan. 2010) Those of metamorphic origin include examples such as those developed in limestones by contact metamorphism (tremolite) and those formed by the alteration of other ferromagnesian minerals (hornblende). Pseudomorphs of amphibole after pyroxene are known as uralite.

History and Etymology

The name amphibole was used by RJ Haüy to include tremolite, actinolite, tourmaline and hornblende. The group was so named by Haüy in allusion to the protean variety, in composition and appearance, assumed by its minerals. This term has since been applied to the whole group. Numerous sub-species and varieties are distinguished, the more important of which are tabulated below in two series. The formulae of each will be seen to be built on the general double-chain silicate formula RSi_4O_{11}.

Mineral Species

Chemical Formulae

Orthorhombic series:

- Anthophyllite $(Mg,Fe)_7Si_8O_{22}(OH)_2$
- Holmquistite $Li_2Mg_3Al_2Si_8O_{22}(OH)_2$.

Monoclinic series:

- Tremolite $Ca_2Mg_5Si_8O_{22}(OH)_2$
- Actinolite $Ca_2(Mg,Fe)_5Si_8O_{22}(OH)_2$
- Cummingtonite $Fe_2Mg_5Si_8O_{22}(OH)_2$
- Grunerite $Fe_7Si_8O_{22}(OH)_2$
- Hornblende $Ca_2(Mg,Fe,Al)_5(Al,Si)_8O_{22}(OH)_2$

- Glaucophane $Na_2(Mg,Fe)_3Al_2Si_8O_{22}(OH)_2$
- Riebeckite (or Crocidolite) $Na_2Fe^{2+}{}_3Fe^{3+}{}_2Si_8O_{22}(OH)_2$
- Arfvedsonite $Na_3Fe^{2+}{}_4Fe^{3+}Si_8O_{22}(OH)_2$
- Richterite $Na_2Ca(Mg,Fe)_5Si_8O_{22}(OH)_2$
- Pargasite $NaCa_2Mg_3Fe^{2+}Si_6Al_3O_{22}(OH)_2$
- Winchite $(CaNa)Mg_4(Al,Fe^{3+})Si_8O_{22}(OH)_2$.

Descriptions

On account of the wide variations in chemical composition, the different members vary considerably in properties and general appearance.

Anthophyllite occurs as brownish, fibrous or lamellar masses with hornblende in mica-schist at Kongsberg in Norway and some other localities. An aluminous related species is known as gedrite and a deep green Russian variety containing little iron as kupfferite.

Hornblende is an important constituent of many igneous rocks. It is also an important constituent of amphibolites formed by metamorphism of basalt.

Actinolite is an important and common member of the monoclinic series, forming radiating groups of acicular crystals of a bright green or greyish-green colour. It occurs frequently as a constituent of greenschists. The name is a translation of the old German word *Strahlstein* (radiated stone).

Glaucophane, crocidolite, riebeckite and arfvedsonite form a somewhat special group of alkali-amphiboles. The first two are blue fibrous minerals, with glaucophane occurring in blueschists and crocidolite (blue asbestos) in ironstone formations, both resulting from dynamo-metamorphic processes. The latter two are dark green minerals, which occur as original constituents of igneous rocks rich in sodium, such as nepheline-syenite and phonolite.

Pargasite is a rare magnesium-rich amphibole with essential sodium, usually found in ultramafic rocks. For instance, it occurs in uncommon mantle xenoliths, carried up by kimberlite. It is hard, dense, black and usually idiomorphic, with a red-brown pleochroism in petrographic thin section.

Analcime

Analcime or analcite is a white, grey, or colourless tectosilicate mineral. Analcime consists of hydrated sodium aluminium silicate in cubic crystalline form. Its chemical formula is $NaAlSi_2O_6 \cdot H_2O$. Minor amounts of potassium and calcium substitute for sodium. A silver-bearing synthetic variety also exists (Ag-analcite).

Analcime is usually classified as a zeolite mineral, but structurally and chemically it is more similar to the feldspathoids. Analcime occurs as a primary mineral in analcime basalt and other alkaline igneous rocks. It also occurs as cavity and vesicle fillings associated with prehnite, calcite, and zeolites.

Locations include the Cyclopean Islands east off Sicily and near Trentino in northern Italy; Victoria in Australia; Kerguelen Island in the Indian Ocean; in the Lake Superior copper district of Michigan, Bergen Hill, New Jersey, Golden, Colorado, and at Searles Lake, California in the United States; and at Cape Blomidon, Nova Scotia and Mont Saint-Hilaire, Quebec in Canada; and in Iceland.

Anapaite

Anapaite is a calcium iron phosphate mineral, $Ca_2Fe^{2+}(PO)_4]_2 \cdot 4H_2O$. It is a mineral that typically occurs in cavities in fossil bearing sedimentary rocks. It is also found in phosphate bearing iron ores.

The spectacular environment that anapaite usually grows in makes it a very popular mineral for collectors. Named after the type locality Anapa, Taman Peninsula, Russia. Some of the best localities are Kertch (Crimea, Ukraine) and Bellaver de Cerdena (Gerona, Spain).

Anatase

Anatase is one of the three mineral forms of titanium dioxide, the other two being brookite and rutile. It is always found as small, isolated and sharply developed crystals, and like rutile, a more commonly occurring modification of titanium dioxide, it crystallizes in the tetragonal system; but, although the degree of symmetry is the same for both, there is no relation between the interfacial angles of the two minerals, except in the prism-zone of 45° and 90°. The common pyramid of anatase, parallel to the faces of which there are perfect cleavages, has an angle over the polar edge of 82°9', the corresponding angle of rutile being 56°52½'. It was on account of this steeper pyramid of anatase that the mineral was named, by René Just Haüy in 1801, from the Greek *anatasis*, "extension", the vertical axis of the crystals being longer than in rutile. There are also important differences between the physical characters of anatase and rutile: the former is less hard (5.5–6 vs. 6-6.5 Mohs) and dense (specific gravity about 3.9

vs. 4.2). Also, anatase is optically negative whereas rutile is positive, and its luster is even more strongly adamantine or metallic-adamantine than that of rutile.

Crystal Habit

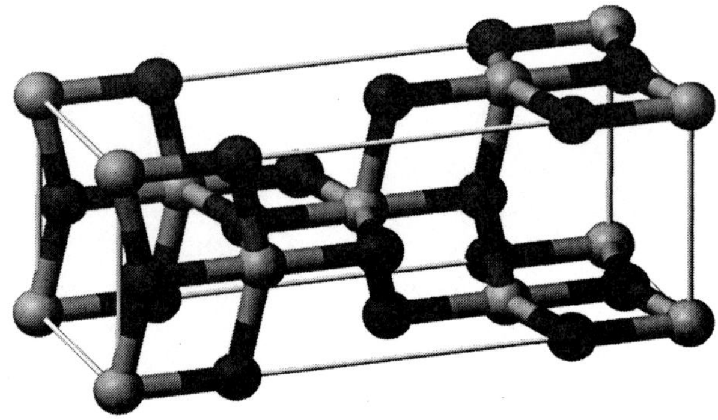

Figure 7 : Crystal structure of anatase

Two types or habits of anatase crystals may be distinguished. The commoner occurs as simple acute double pyramids with an indigo-blue to black colour and steely luster.

Crystals of this kind are abundant at Le Bourg-d'Oisans in Dauphiné, where they are associated with rock-crystal, feldspar, and axinite in crevices in granite and mica-schist.

Similar crystals, but of microscopic size, are widely distributed in sedimentary rocks, such as sandstones, clays, and slates, from which they may be separated by washing away the lighter constituents of the powdered rock.

Crystals of the second type have numerous pyramidal faces developed, and they are usually flatter or sometimes prismatic in habit; the colour is honey-yellow to brown. Such crystals closely resemble xenotime in appearance and, indeed, were for a long time supposed to belong to this species, the special name *wiserine* being applied to them.

They occur attached to the walls of crevices in the gneisses of the Alps, the Binnenthal near Brig in canton Valais, Switzerland, being a well-known locality. Naturally occurring pseudomorphs of rutile after anatase are also known.

While anatase is not an equilibrium phase of TiO_2, it is kinetically stabilized. At temperatures between 550 and about 1000 °C, anatase transforms to the equilibrium rutile phase, increasing its specific gravity to 4.2. The temperature of this transformation strongly depends on the impurities or dopants present in the material as well as on the morphology of the sample.

Synthetic Anatase

Crystals of anatase can be prepared in laboratories by chemical methods such as sol-gel processing.

Examples include controlled hydrolysis of titanium tetrachloride ($TiCl_4$) or titanium alkoxides. Such synthetic anatases are of considerable interest for photocatalysis and for Graetzel-cell photovoltaics.

Alternate and Obsolete Names

Another name commonly in use for this mineral is *octahedrite*, a name which, indeed, is earlier than anatase, and given because of the common (acute) octahedral habit of the crystals. Other names, now obsolete, are oisanite and dauphinite, from the well-known French locality.

Andalusite

Andalusite is an aluminium nesosilicate mineral with the chemical formula Al_2SiO_5.

The variety chiastolite commonly contains dark inclusions of carbon or clay which form a checker-board pattern when shown in cross-section.

A clear variety first found in Andalusia, Spain can be cut into an interesting gemstone. Faceted andalusite stones give a play of red, green, and yellow colours that resembles a muted form of iridescence, although the colours are actually the result of unusually strong pleochroism. It is associated with mica schist which increases alkali content in ultimate product and so it has not been exploited economically so far.

Occurrence

Andalusite is a common regional metamorphic mineral which forms under low pressure and moderate to high temperatures. Called Lapis Crucifer in ancient texts. The minerals kyanite and sillimanite are polymorphs of andalusite, each occurring under different temperature-pressure regimes and are therefore rarely found together in the same rock. Because of this the three minerals are a useful tool to help identify the pressure-temperature paths of the host rock in which they are found.

It was first described and named after the type locality in the Ronda Massif, Málaga, Andalusia, Spain in 1789.

Andesine

Andesine is a silicate mineral, a member of the plagioclase feldspar solid solution series. Its chemical formula is $(Ca, Na)(Al, Si)_4O_8$, where Ca/(Ca + Na) (% Anorthite) is between 30%-50%.

The formula may be written as $Na_{0.7\text{-}0.5}Ca_{0.3\text{-}0.5}Al_{1.3\text{-}1.5}Si_{2.7\text{-}2.5}O_8$.

The plagioclase feldspars are a continuous solid solution series and as such the accurate identification of individual members requires detailed optical study, chemical analysis or density measurements. Refractive indices and specific gravity increase directly with calcium content.

Name and Discovery

Andesine was first described in 1841 for an occurrence in the Marmato mine, Marmato, Cauca, Chocó Department, Colombia. The name is for the Andes due to its abundance in the andesite lavas in those mountains.

Occurrence

Andesine occurs in intermediate igneous rocks such as diorite, syenite and andesite. It characteristically occurs in metamorphic rocks of granulite to amphibolite facies commonly exhibiting antiperthite texture.

It also occurs as detrital grains in sedimentary rocks. It is commonly associated with quartz, potassium feldspar, biotite, hornblende and magnetite.

Andradite

Andradite is a species of the garnet group. It is a nesosilicate, with formula $Ca_3Fe_2Si_3O_{12}$.

Andradite includes three varieties:

- *Melanite*: Black in colour, referred to as "titanian andradite".
- Demantoid: Vivid green in colour, one of the most valuable and rare stones in the gemological world.
- *Topazolite*: Yellow-green in colour and sometimes of high enough quality to be cut into a faceted gemstone.

It was first described in 1868 for an occurrence in Drammen, Buskerud, Norway. Andradite was named after the Brazilian mineralogist José Bonifácio de Andrade e Silva (1763–1838).

Occurrence

It occurs in skarns developed in contact metamorphosed impure limestones or calcic igneous rocks; in chlorite schists and serpentinites and in alkalic igneous rocks (typically titaniferous). Associated minerals include vesuvianite, chlorite, epidote, spinel, calcite, dolomite and magnetite.

It is found in Italy, the Ural Mountains of Russia, Arizona and California and in Dnipropetrovsk Oblast in Ukraine.

As the other garnets andradite crystallizes in the cubic space group [[Ia3d]], with unit-cell parameter of 12.051 Å at 100 K.

The spin structure of andradite contains two mutually canted equivalent antiferromagnetic sublattices below the Néel temperature (T_N=11 K).

Anglesite

Anglesite is a lead sulfate mineral with the chemical formula $PbSO_4$. It occurs as an oxidation product of primary lead sulfide ore, galena.

Anglesite occurs as prismatic orthorhombic crystals and earthy masses, and is isomorphous with barite and celestine. It contains 74% of lead by mass and therefore has a high specific gravity of 6.3. Colour is white or gray with pale yellow streaks. It may be dark gray if impure.

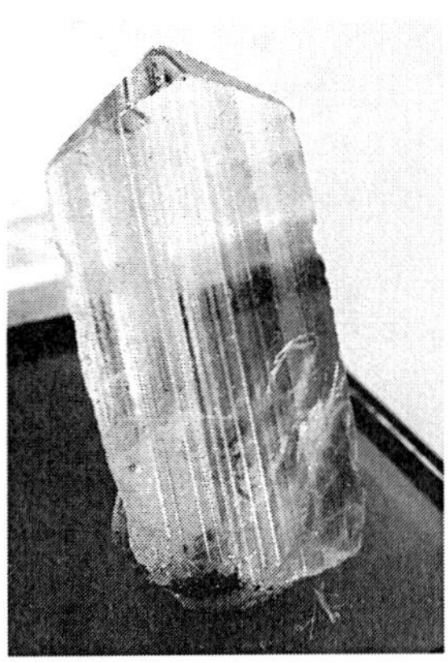

Figure 8 : Anglesite Crystal from Touissit District, Morocco (size: 2.8×1.6×0.5 cm)

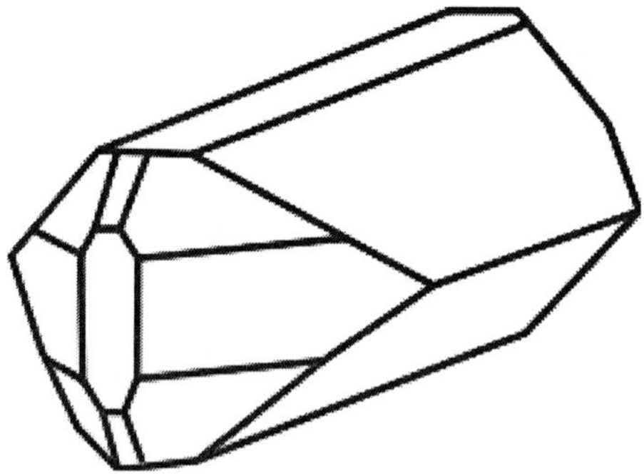

Figure 9 : Anglesite Diagram Illustrating its Orthorhombic Crystalline form

It was first recognized as a mineral species by William Withering in 1783, who discovered it in the Parys copper-mine in Anglesey; the name anglesite, from this locality, was given by F. S. Beudant in 1832. The crystals from Anglesey, which were formerly found abundantly on a matrix of dull limonite, are small in size and simple in form, being usually bounded by four faces of a prism and four faces of a dome; they are brownish-yellow in colour owing to a stain of limonite. Crystals from some other localities, notably from Monteponi in Sardinia, are transparent and colourless, possessed of a brilliant adamantine lustre, and usually modified by numerous bright faces. The variety of combinations and habits presented by the crystals is very extensive, nearly two hundred distinct forms being figured by V. von Lang in his monograph of the species; without measurement of the angles the crystals are frequently difficult to decipher. There are distinct cleavages parallel to the faces of the prism (110) and the basal plane (001), but these are not so well developed as in the isomorphous minerals barite and celestite.

Anglesite is a mineral of secondary origin, having been formed by the oxidation of galena in the upper parts of mineral lodes where these have been affected by weathering processes. At Monteponi the crystals encrust cavities in glistening granular galena; and from Leadhills, in Scotland, pseudomorphs of anglesite after galena are known. At most localities it is found as isolated crystals in the lead-bearing lodes, but at some places, in Australia and Mexico, it occurs as large masses, and is then mined as an ore of lead.

Anhydrite

Anhydrite is a mineral – anhydrous calcium sulfate, $CaSO_4$. It is in the orthorhombic crystal system, with three directions of perfect cleavage parallel to the three planes of symmetry. It is not isomorphous with the orthorhombic barium (baryte) and strontium (celestine) sulfates, as might be expected from the chemical formulas. Distinctly developed crystals are somewhat rare, the mineral usually presenting the form of cleavage masses. The hardness is 3.5 and the specific gravity 2.9. The colour is white, sometimes greyish, bluish or purple. On the best developed of the three cleavages the lustre is pearly, on other surfaces it is vitreous. When exposed to water, anhydrite readily transforms to the more commonly occurring gypsum, ($CaSO_4 \cdot 2H_2O$) by the absorption of water. This transformation is reversible, with gypsum or calcium sulfate hemihydrate forming anhydrite by heating to ~200°C under normal atmospheric conditions. Anhydrite is commonly associated with calcite, halite, and sulfides such as galena, chalcopyrite, molybdenite and pyrite in vein deposits.

Occurrence

Anhydrite is most frequently found in evaporite deposits with gypsum; it was, for instance, first discovered, in 1794, in a salt mine near Hall in Tirol. In this occurrence depth is critical since nearer the surface anhydrite has been altered to gypsum by absorption of circulating ground water.

From an aqueous solution calcium sulfate is deposited as crystals of gypsum, but when the solution contains an excess of sodium or potassium chloride anhydrite is deposited if temperature is above 40 °C. This is one of the several methods by which the mineral has been prepared artificially, and is identical with its mode of origin in nature, the mineral is common in salt basins.

Tidal Flat Nodules

Anhydrite occurs in a tidal flat environment in the Persian Gulf sabkhas as massive diagenetic replacement nodules. Cross sections of these nodular masses have a *netted* appearance and have been referred to as *chicken wire anhydrite*. Nodular anhydrite occurs as replacement of gypsum in a variety of sedimentary depositional environments.

Salt Dome Cap Rocks

Massive amounts of anhydrite occur when salt domes form a caprock. Anhydrite is 1–3% of the salt in salt domes and is generally left as a cap at the top of the salt when the halite is removed by pore waters. The typical cap rock is a salt, topped by a layer of anhydrite, topped by patches of gypsum, topped by a layer of calcite. Interaction with oil can reduce SO_4 creating calcite, water, and hydrogen sulfide (H_2S).

Naming History

The name anhydrite was given by A. G. Werner in 1804, because of the absence of water of crystallization, as contrasted with the presence of water in gypsum. Some obsolete names for the species are muriacite and karstenite; the former, an earlier name, being given under the impression that the substance was a chloride (muriate). A peculiar variety occurring as contorted concretionary masses is known as tripe-stone, and a scaly granular variety, from Volpino, near Bergamo, in Lombardy, as vulpinite; the latter is cut and polished for ornamental purposes.

Ankerite

Ankerite is a calcium, iron, magnesium, manganese carbonate mineral of the group of rhombohedral carbonates with formula: $Ca(Fe, Mg, Mn)(CO_3)_2$. In composition it is closely related to dolomite, but differs from this in having magnesium replaced by varying amounts of iron(II) and manganese.

The crystallographic and physical characters resemble those of dolomite and siderite. The angle between the perfect rhombohedral cleavages is 73° 48', the hardness is 3.5 to 4, and the specific gravity is 2.9 to 3.1. The colour is white, grey or reddish to yellowish brown.

Ankerite occurs with siderite in deposits of iron-ore. It is one of the minerals of the dolomite-siderite series, to which the terms brown-spar, pearl-spar and bitter-spar have been historically loosely applied.

Ankerite can result from hydrothermal or direct ground water precipitation. It can also be the result of metamorphic recrystallization of iron-rich sedimentary rocks. It is often found as a gangue mineral associated with gold and a variety of sulfide minerals in ore deposits.

It was first recognized as a distinct species by W. von Haidinger in 1825, and named by him after Matthias Joseph Anker (1771-1843) of Styria, an Austrian mineralogist.

Annabergite

Annabergite is an arsenate mineral consisting of a hydrous nickel arsenate, $Ni_3(AsO_4)_2{\cdot}8H_2O$, crystallizing in the monoclinic system and isomorphous with vivianite and erythrite. Crystals are minute and capillary and rarely met with, the mineral occurring usually as soft earthy masses and encrustations. A fine apple-green colour is its characteristic feature. It was long known (since 1758) under the name nickel ochre; the name annabergite was proposed by H. J. Brooke and W H. Miller in 1852, from Annaberg in Saxony, one of the localities of the mineral. It occurs with ores of nickel, of which it is a product of alteration. A variety, from Creetown in Kirkcudbrightshire, in which a portion of the nickel is replaced by calcium, has been called *dudgeonite*, after P. Dudgeon, who found it.

Closely related is *cabrerite* wherein some of the nickel is replaced by magnesium. It is named for Sierra Cabrera in Spain where it was originally found.

Anorthite

Anorthite is the calcium endmember of plagioclase feldspar. Plagioclase is an abundant mineral in the Earth's crust. The formula of pure anorthite is $CaAl_2Si_2O_8$.

Anorthoclase

Figure 10 : An Anorthoclase crystal from Mount Erebus, Antarctica

The mineral anorthoclase ($(Na,K)AlSi_3O_8$) is a crystalline solid solution in the alkali feldspar series, in which the sodium-aluminium silicate member exists in larger proportion. It typically consists of between 10 and 36 percent of $KAlSi_3O_8$ and between 64 and 90 percent of $NaAlSi_3O_8$.

Anthophyllite

Anthophyllite is an amphibole mineral: $(Mg, Fe)_7Si_8O_{22}(OH)_2$, magnesium iron inosilicate hydroxide. Anthophyllite is polymorphic with cummingtonite. Some forms of anthophyllite are lamellar or fibrous and are used as asbestos. The name is derived from the Latin word *anthophyllum*, meaning *clove*, an allusion to the most common colour of the mineral.

Occurrence

Anthophylite is the product of metamorphism of magnesium-rich rocks, especially ultrabasic igneous rocks and impure dolomitic shales. It also forms as a retrograde product rimming relict orthopyroxenes and olivine, and as an accessory mineral in cordierite-bearing gneisses and schists. Anthophyllite also occurs as a retrograde metamorphic mineral derived from ultramafic rocks along with serpentinite. Geographically, it occurs in Pennsylvania, southwestern New Hampshire, central Massachusetts, Franklin, North Carolina, and in the Gravelly Range and Tobacco Root Mountains of southwest Montana.

Occurrence in Ultramafic Rocks

Anthophyllite is formed by the breakdown of talc in ultramafic rocks in the presence of water and carbon dioxide as a prograde metamorphic reaction. The partial pressure of carbon dioxide (XCO_2) in aqueous solution favours production of anthophyllite. Higher partial pressures of CO_2 reduces the temperature of the *anthophyllite-in* isograd.

Figure 11 : Anthophyllite in Serpentinised Komatiite, Maggie Hays Ni Mine, Western Australia

Ultramafic rocks in purely hydrous, CO_2-free environments will tend to form serpentinite-antigorite-brucite-tremolite assemblages (dependent on MgO content) or at amphibolite to granulite metamorphic grade, metamorphic pyroxene or olivine. Thus, metamorphic assemblages of ultramafic rocks containing anthophyllite are indicative of at least greenschist facies metamorphism in the presence of carbon dioxide bearing metamorphic fluids.

The typical metamorphic assemblage reactions for low-magnesian (<25% MgO) and high-magnesian (>25% MgO) ultramafic rocks are;

- Olivine + Tremolite + Talc → Olivine + Tremolite + Anthophyllite (low MgO, >550°C, XCO2 <0.6)
- Talc + Tremolite + Magnesite →Tremolite + Anthophyllite + Magnesite (High MgO, >500°C, XCO2 >0.6)
- Talc + Magnesite + Tremolite →Anthophyllite + Tremolite + Magnesite (Low MgO, >500°C, XCO2 >0.6).

Retrogressive anthophyllite is relatively rare in ultramafic rocks and is usually poorly developed due to the lower energy state available for metamorphic reactions to progress and also the general dehydration of rock masses during metamorphism. Similarly, the need for substantial components of carbon dioxide in metamorphic fluid restricts the appearance of anthophyllite as a retrograde mineral. The usual metamorphic assemblage of retrograde-altered ultramafic rocks is thus usually a serpentinite or talc-magnesite assemblage.

Retrograde anthophyllite is present most usually in shear zones where fracturing and shearing of the rocks provides a conduit for carbonated fluids during retrogression.

Chapter 6

Serpentine Group

The serpentine group describes a group of common rock-forming hydrous magnesium iron phyllosilicate ($(Mg, Fe)_3Si_2O_5(OH)_4$) minerals; they may contain minor amounts of other elements including chromium, manganese, cobalt and nickel. In mineralogy and gemology, serpentine may refer to any of 20 varieties belonging to the serpentine group. Owing to admixture, these varieties are not always easy to individualize, and distinctions are not usually made. There are three important mineral polymorphs of serpentine: antigorite, chrysotile and lizardite.

Figure 1 : Serpentine

Many types of serpentine have been used for jewellery and hardstone carving, sometimes under the name *false jade* or *Teton jade*.

"Their olive green colour and smooth or scaly appearance is the basis of the name from the Latin *serpentinus*, meaning "serpent rock," according to Best (2003). They have their origins in metamorphic alterations of peridotite and pyroxene. Serpentines may also pseudomorphously replace other magnesium silicates. Alterations may be incomplete, causing physical properties of serpentines to vary widely. Where they form a significant part of the land surface, the soil is unusually high in clay.

Antigorite is the polymorph of serpentine that most commonly forms during metamorphism of wet ultramafic rocks and is stable at the highest temperatures—to over 600 °C at depths of 60 km or so. In contrast, lizardite and chrysotile typically form near the Earth's surface and break down at relatively low temperatures, probably well below 400 °C. It has been suggested that chrysotile is never stable relative to either of the other two serpentine polymorphs.

Samples of the oceanic crust and uppermost mantle from ocean basins document that ultramafic rocks there commonly contain abundant serpentine. Antigorite contains water in its structure, about 13 percent by weight. Hence, antigorite may play an important role in the transport of water into the earth in subduction zones and in the subsequent release of water to create magmas in island arcs, and some of the water may be carried to yet greater depths.

Soils derived from serpentine are toxic to many plants, because of high levels of nickel, chromium, and cobalt; growth of many plants is also inhibited by low levels of potassium and phosphorus and a low ratio of calcium/magnesium. The flora is generally very distinctive, with specialised, slow-growing species. Areas of serpentine-derived soil will show as strips of shrubland and open, scattered small trees (often conifers) within otherwise forested areas; these areas are called serpentine barrens.

Most serpentines are opaque to translucent, light (specific gravity between 2.2–2.9), soft (hardness 2.5–4), infusible and susceptible to acids. All are microcrystalline and massive in habit, never being found as single crystals. Luster may be vitreous, greasy or silky. Colours range from white to grey, yellow to green, and brown to black, and are often splotchy or veined.

Many are intergrown with other minerals, such as calcite and dolomite. Occurrence is worldwide; New Caledonia, Canada (Quebec), USA (northern California, Rhode Island, Connecticut, Massachusetts, Maryland and southern Pennsylvania), Afghanistan, Britain (Cornwall and Ireland), Greece (Thessaly), China, France, Korea, Austria (Styria and Carinthia), India (Assam, and Manipur), Myanmar (Burma), New Zealand, Norway and Italy are notable localities.

Rock composed primarily of these minerals is called serpentinite. Serpentines find use in industry for a number of purposes, such as railway ballasts, building materials, and the asbestiform types find use as thermal and electrical insulation (chrysotile asbestos). The asbestos content can be released to the air when serpentine is excavated and if it is used as a road surface, forming a long term health hazard by breathing.

Asbestos from serpentine can also appear at low levels in water supplies through normal weathering processes, but there is as yet no identified health hazard associated with use or ingestion. In its natural state, some forms of serpentine react with carbon dioxide and re-release oxygen into the atmosphere.

The more attractive and durable varieties (all of antigorite) are termed "noble" or "precious" serpentine and are used extensively as gems and in ornamental carvings.

The town of Bhera in the historic Punjab province of the Indian subcontinent was known for centuries for finishing a relatively pure form of green serpentine obtained from quarries in Afghanistan into lapidary work, cups, ornamental sword hilts, and dagger handles. This high-grade serpentine ore was known as *sang-i-yashm* or to the English, *false jade*, and was used for generations by Indian craftsmen for lapidary work. It is easily carved, taking a good polish, and is said to have a pleasingly greasy feel. Less valuable serpentine ores of varying hardness and clarity are also sometimes dyed to imitate jade. Misleading synonyms for this material include "Suzhou jade", "Styrian jade", and "New jade".

New Caledonian serpentine is particularly rich in nickel, and is the source of most of the world's nickel ore. The Mâori of New Zealand once carved beautiful objects from local serpentine, which they called *tangiwai*, meaning "tears".

The *lapis atracius* of the Romans, now known as verde antique, or verde antico, is a serpentinite breccia popular as a decorative facing stone. In classical times it was mined at Casambala, Thessaly, Greece. Serpentinite marbles are also widely used: Green Connemara marble (or Irish green marble) from Connemara, Ireland (and many other sources), and red Rosso di Levanto marble from Italy. Use is limited to indoor settings as serpentinites do not weather well.

Antigorite

Lamellated antigorite occurs in tough, pleated masses. It is usually dark green in colour, but may also be yellowish, gray, brown or black. It has a hardness of 3.5–4 and its lustre is greasy. The monoclinic crystals show micaceous cleavage and fuse with difficulty. Antigorite is named after its type locality, the Valle di Antigorio in Italy.

Bowenite is an especially hard serpentine (5.5) of a light to dark apple green colour, often mottled with cloudy white patches and darker veining. It is the serpentine most frequently encountered in carving and jewellery. The name retinalite is sometimes applied to yellow bowenite. The New Zealand material is called tangiwai. Although not an official species, bowenite is the state mineral of Rhode Island: this is also the variety's type locality. A bowenite cabochon featured as part of the "Our Mineral Heritage Brooch", was presented to First Lady Mrs. Lady Bird Johnson in 1967. Williamsite is a local varietal name for antigorite that is oil-green with black crystals of chromite or magnetite often included. Somewhat resembling fine jade, williamsite is cut into cabochons and beads. It is found mainly in Maryland and Pennsylvania, USA.

Gymnite is an amorphous form of antigorite. It was originally found in the Bare Hills, Maryland, and is named from the Greek, gymnos meaning bare or naked.

Lizardite

Extremely fine-grained, scaly lizardite (also called orthoantigorite) comprises much of the serpentine present in serpentine marbles. It is triclinic, has one direction of perfect cleavage, and may be white, yellow or green. Lizardite is translucent, soft (hardness 2.5) and has an average specific gravity of 2.57. It can be pseudomorphous after enstatite, olivine or pyroxene, in which case the name bastite is sometimes applied. Bastite may have a silky lustre. Lizardite is named after its type locality on the Lizard Peninsula, Cornwall, UK. It is worked by local artisans into various trinkets which are sold to tourists.

State Emblem

In 1965 the California Legislature designated serpentine (the mineral) as "the official State Rock and lithologic emblem." In May, 2010, State Senator Gloria J. Romero proposed legislation (SB 624) that would remove serpentine as the state rock and lithologic emblem and would leave the state rock unspecified.

Antimony

Antimony is a toxic chemical element with the symbol Sb and an atomic number of 51. A lustrous grey metalloid, it is found in nature mainly as the sulfide mineral stibnite (Sb_2S_3). Although the use of antimony is limited by its toxicity, its compounds have been of fundamental value in chemistry – a prominent example being the development of superacids derived from antimony pentafluoride. Antimony compounds are prominent fire retardants found in many commercial and domestic products. Certain alloys are valuable for use in solders and ball bearings. An emerging application is the use of antimony in microelectronics.

History

An artefact, said to be part of a vase, made of antimony dating to about 3000 BC was found at Tello, Chaldea (part of present-day Iraq), and a copper object plated with antimony dating between 2500 BC and 2200 BC has been found in Egypt.

One contemporary (Austen, at a lecture by Herbert Gladstone, published in 1892) was reported to comment that "we only know of antimony at the present day as a highly brittle and crystalline metal, which could hardly be fashioned into a useful vase, and therefore this remarkable 'find' must represent the lost art of rendering antimony malleable." However, Moorey was unconvinced that the artefact was indeed a vase, mentioning that Selimkhanov, after his analysis of the Telloh object (published in 1975), "attempted to relate the metal to Transcaucasian natural antimony" (i.e. native metal) and that "the antimony objects from Transcaucasia are all small personal ornaments." This weakens the evidence for a lost art "of rendering antimony malleable."

The first European description of a procedure for isolating antimony is in the book *De la pirotechnia* of 1540 by Vannoccio Biringuccio. This book predates the more famous 1556 book by Agricola, *De re metallica*, even though Agricola has been often incorrectly credited with the discovery of metallic antimony. A text describing the preparation of metallic antimony that was published in Germany in 1604 purported to date from the early fifteenth century, and if authentic it would predate Biringuccio. The book, written in Latin, was called "Currus Triumphalis Antimonii" (The Triumphal Chariot of Antimony), and its putative author was a certain Benedictine monk, writing under the name Basilius Valentinus. An English translation of the "Currus Triumphalis" appeared in English in 1660, under the title The Triumphant Chariot of Antimony. The work remains of great interest, chiefly because it documents how followers of the renegade German physician, Philippus Theophrastus Paracelsus von Hohenheim (of whom Thölde was one), came to associate the practice of alchemy with the preparation of chemical medicines.

Pure antimony was well known to Jâbir ibn Hayyân, sometimes called "the Father of Chemistry", in the 8th century. Here there is still an open controversy: Marcellin Berthelot, who translated a number of Jâbir's books, stated that antimony is never mentioned in them, but other authors claim that Berthelot translated only some of the less important books, while the more interesting ones (some of which might describe antimony) are not yet translated, and their content is completely unknown.

The first natural occurrence of pure antimony ('native antimony') in the Earth's crust was described by the Swedish scientist and local mine district engineer Anton von Swab in 1783. The type-sample was collected from the Sala Silver Mine in the Bergslagen mining district of Sala, Västmanland, Sweden.

Etymology

The ancient words for antimony mostly have, as their chief meaning, kohl, the sulfide of antimony. Pliny the Elder, however, distinguishes between male and female forms of antimony; his male form is probably the sulfide, while the female form, which is superior, heavier, and less friable, is probably native metallic antimony.

The Egyptians called antimony *m¿dmt*; in hieroglyphs, the vowels are uncertain, but there is an Arabic tradition that the word is *mesdemet*. The Greek word, *stimmi*, is probably a loan word from Arabic or Egyptian *sdm*, and is used by the Attic tragic poets of the 5th century BC; later Greeks also used *stibi*, as did Celsus and Pliny, writing in Latin, in the first century AD. Pliny also gives the names *stimi* [*sic*], *larbaris*, alabaster, and the "very common" *platyophthalmos*, "wide-eye" (from the effect of the cosmetic). Later Latin authors adapted the word to Latin as *stibium*. The Arabic word for the substance, as opposed to the cosmetic, can appear as *ithmid, athmoud, othmod*, or *uthmod*. Littré suggests the first form, which is the earliest, derives from *stimmida*, (one) accusative for *stimmi*.

The use of Sb as the standard chemical symbol for antimony is due to the 18th century chemical pioneer, Jöns Jakob Berzelius, who used this abbreviation of the name *stibium*. The medieval Latin form, from which the modern languages and late Byzantine Greek, take their names, is *antimonium*. The origin of this is uncertain; all suggestions have some difficulty either of form or interpretation. The popular etymology, from *anti-monachos* or French *antimoine*, still has adherents; this would mean "monk-killer", and is explained by many early alchemists being monks, and antimony being poisonous. So does the hypothetical Greek word *antimonos*, "against one", explained as "not found as metal", or "not found unalloyed". Lippmann conjectured a Greek word, *anthemonion*, which would mean "floret", and he cites several examples of related Greek words (but not that one) which describe chemical or biological efflorescence.

The early uses of *antimonium* include the translations, in 1050–1100, by Constantine the African of Arabic medical treatises. Several authorities believe that *antimonium* is a scribal corruption of some Arabic form; Meyerhof derives it from *ithmid*; other possibilities include *Athimar*, the Arabic name of the *metal*, and a hypothetical *as-stimmi*, derived from or parallel to the Greek.

Antitaenite

Antitaenite is a meteoritic metal alloy mineral composed of iron and nickel, 20-40% Ni (and traces of other elements) that has a face centred cubic crystal structure. Its existence as a new mineral species occurring in both iron meteorites and in chondrites was first recognized in 1995. There are three other known Fe-Ni meterotic minerals: kamacite, taenite, and tetrataenite. The pair of minerals antitaenite and taenite constitute the first example in nature of two minerals that have the same crystal structure (face centred cubic) and can have the same chemical composition (same proportions of Fe and Ni)-they differ in their electronic structures: taenite is a high magnetic moment alloy whereas antitaenite is a low magnetic moment alloy. This unique difference in electronic structure was first established in 1999 and arises from a high-magnetic-moment to low-magnetic-moment transition occurring in the Fe-Ni bi-metallic alloy series. The same electronic structure transition is believed to be a causal factor in Invar behaviour.

Antlerite

Antlerite is a greenish hydrous copper sulfate mineral, with the formula $Cu_3(SO_4)(OH)_4$. It occurs in tabular, acicular, or fibrous crystals with a vitreous luster. Originally believed to be a rare mineral, antlerite was found to be the primary ore of the oxidised zones in several copper mines across the world, including the Chuquicamata mine in Chile, and the Antler mine in Arizona, USA from which it takes its name. It is chemically and optically similar in many respects to other copper minerals such as malachite and brochantite, though it can be distinguished from the former by a lack of effervescence in hydrochloric acid.

Barium

Barium is a chemical element with the symbol Ba and atomic number 56. It is the fifth element in Group 2, a soft silvery metallic alkaline earth metal. Barium is never found in nature in its pure form due to its reactivity with air. Its oxide is historically known as baryta but it reacts with water and carbon dioxide and is not found as a mineral. The most common naturally occurring minerals are the very insoluble barium sulfate, $BaSO_4$ (barite), and barium carbonate, $BaCO_3$ (witherite). Barium's name originates from Greek *barys*, meaning "heavy", describing the high density of some common barium-containing ores.

Barium has few industrial applications, but the metal has been historically used to scavenge air in vacuum tubes. Barium compounds impart a green colour to flames and have been used in fireworks. Barium sulfate is used for its density, insolubility, and X-ray opacity. It is used as an insoluble heavy mud-like paste when drilling oil wells, and in purer form, as an X-ray radiocontrast agent for imaging the human gastrointestinal tract. Soluble barium compounds are poisonous due to release of the soluble barium ion, and have been used as rodenticides. New uses for barium continue to be sought. It is a component of some "high temperature" YBCO superconductors, and electroceramics.

Characteristics

Figure 2: Barite

Physical Properties

Barium is a soft, silvery white alkali earth metal, which quickly oxidizes in air. It crystallizes in body centred cubic lattices. It burns with a green to pale green flame, resulting from emission at 524.2 and 513.7 nm. Its simple compounds are notable for their relatively high (for an alkaline earth element) specific gravity. This high density is true of the most common barium-bearing mineral, barite ($BaSO_4$), also called 'heavy spar' due to the high density (4.5 g/cm^3).

Chemical Properties

Barium, as for other alkali earth (group II) metals, is highly reducing. It reacts exothermically with oxygen at room temperature to form barium oxide and peroxide. Because of its sensitivity to air, samples are generally stored under protective oils. The reaction is violent if barium is powdered. The metal is readily attacked in most acids, with the notable exception of sulfuric acid, as passivation stops the reaction by forming the insoluble barium sulfate. It also reacts violently with water according to the reaction:

$$Ba + 2\ H_2O \rightarrow Ba(OH)_2 + H_2\uparrow$$

Barium combines with several metals, including aluminium, zinc, lead and tin, forming intermetallic phases and alloys.

Isotopes

Naturally occurring barium is a mix of seven stable isotopes, the most abundant being ^{138}Ba (71.7 %). 22 isotopes are known, but most of these are highly radioactive and have half-lives in the several millisecond to several day range. The only notable exceptions are ^{133}Ba which has a half-life of 10.51 years, and ^{137m}Ba (2.55 minutes). ^{133}Ba is a standard calibrant for gamma-ray detectors in nuclear physics studies.

Occurrence

The abundance of barium is 0.0425 % in the Earth's crust and 13 µg/L in sea water. It occurs in the minerals barite (as the sulfate) and witherite (as the carbonate). Although witherite deposits were mined from the 17th century till 1969 in northern England, for example in the Settlingstones Mine near Newbrough, today nearly all barium is mined as barite.

Large deposits of barite are found in China, Germany, India, Morocco, and in the United States. A rare gem containing barium is known, called benitoite.

Production

Because barium quickly oxidizes in air, it is difficult to obtain the free metal and it is never found free in nature. The metal is primarily found in, and extracted from, barite. Because barite is so insoluble, it cannot be used directly for the preparation of other barium compounds, or barium metal. Instead, the ore is heated with carbon to reduce it to barium sulfide:

$$BaSO_4 + 2\ C \rightarrow BaS + 2\ CO_2$$

The barium sulfide is then hydrolyzed or treated with acids to form other barium compounds, such as the chloride, nitrate, and carbonate. Barium is commercially produced through the electrolysis of molten barium chloride ($BaCl_2$):

$$\text{(cathode) } Ba^{2+} + 2\ e^- \rightarrow Ba$$

$$\text{(anode) } 2\ Cl^- \rightarrow Cl_2 + 2\ e^-$$

Barium metal is also obtained by the reduction of barium oxide with finely divided aluminium at temperatures between 1100 and 1200 °C:

$$4\ BaO + 2\ Al \rightarrow BaO\cdot Al_2O_3 + 3\ Ba$$

The barium vapour is cooled and condensed to give the solid metal, which can be cast into rods or extruded into wires. Being a flammable solid, it is packaged under argon in steel containers or plastic bags.

Compounds

Ba^{2+} is the dominant oxidation state throughout the chemistry of barium. Its properties generally resemble those of other alkaline earth ions such as strontium and calcium. All halides, pseudohalides and chalcogenides are known, usually as colourless solids. The sulfate is famously insoluble. BaO forms a peroxide when heated in air. The oxide is basic and reacts with acids to give salts. Barium reduces oxides, chlorides and sulfides of less active metals. For example:

$$Ba + CdO \rightarrow BaO + Cd$$

$$Ba + ZnCl_2 \rightarrow BaCl_2 + Zn$$

$$3\ Ba + Al_2S_3 \rightarrow 3\ BaS + 2\ Al$$

At elevated temperatures, barium combines with nitrogen and hydrogen to produce the nitride Ba_3N_2 and hydride BaH_2, respectively. When heated with nitrogen and carbon, it forms the cyanide:

$$Ba + N_2 + 2\ C \rightarrow Ba(CN)_2$$

History

Barium's name originates from Greek *barys*, meaning "heavy", describing the density of some common barium-containing ores. Alchemists in the early Middle Ages knew about some barium minerals. Smooth pebble-like stones of mineral barite found in Bologna, Italy were known as "Bologna stones". Witches and alchemists were attracted to them because after exposure to light they would glow for years.

Carl Scheele identified barite as containing a new element in 1774, but could not isolate barium, only barium oxide. Johan Gottlieb Gahn also isolated barium oxide two years later in similar studies. Oxidized barium was at first called barote, by Guyton de Morveau, a name which was changed by Antoine Lavoisier to *baryta.*

Also in the 18th century, English mineralogist William Withering noted a heavy mineral in the lead mines of cumberland, now known to be Witherite. Barium was first isolated by electrolysis of molten barium salts in 1808, by Sir Humphry Davy in England.

Davy, by analogy with calcium named "barium" after baryta, with the "-ium" ending signifying a metallic element. Robert Bunsen and Augustus Matthiessen yielded pure barium by electrolysis of a molten mixture of barium chloride and ammonium chloride.

The production of pure oxygen in the Brin process was a large scale application of barium peroxide before electrolysis and fractionally distill liquefied air became the dominant ways to produce oxygen. In this process the barium oxide reacts at 500–600°C with air to form barium peroxide which decomposes at above 700°C by releasing oxygen.

$$2\ BaO + O_2 \rightleftharpoons 2\ BaO_2$$

Applications

The dominating application of elemental barium is as a scavenger or "getter" removing the last traces of oxygen and other gases in electronic vacuum tubes such as television cathode ray tubes.

An alloy of barium with nickel is commonly used in automobile ignitions.

Applications of Barium Sulfate

Barium sulfate (the mineral barite, $BaSO_4$) is important to the petroleum industry, for example, as drilling mud, a weighting agent in drilling new oil wells. It is also a filler in a variety of products such as rubber. Taking advantage of its opacity to X-rays, the sulfate is used as a radiocontrast agent for X-ray imaging of the digestive system ("barium meals" and "barium enemas"). Lithopone, a pigment that contains barium sulfate and zinc sulfide, is a permanent white that has good covering power, and does not darken when exposed to sulfides.

Applications of other barium compounds

Aside from the sulfate, other compounds of barium find only niche applications. Applications are limited by the toxicity of Ba^{2+} ions (Barium carbonate is a rat poison), which is not a problem for the insoluble $BaSO_4$.

- Barium oxide is used in a coating for the electrodes of fluorescent lamps, which facilitates the release of electrons.
- Barium carbonate is also used in glassmaking. Being a heavy element, barium increases the refractive index and luster of the glass.
- Barium, commonly as barium nitrate, is used to give green colours in fireworks. The species responsible for the brilliant green is barium monochloride; in the absence of a source of chlorine a yellow or "apple" green is produced instead.
- Barium peroxide can be used as a catalyst to start an aluminothermic reaction when welding rail tracks together. It can also be used in green tracer ammunition and as a bleaching agent.
- Barium titanate is a promising electroceramic.
- Barium fluoride is used for optics in infrared applications, since it is transparent from about 0.15 to 12 micrometres.

Precautions

Soluble barium compounds are poisonous. At low doses, barium acts as a muscle stimulant, whereas higher doses affect the nervous system, causing cardiac irregularities, tremors, weakness, anxiety, dyspnea and paralysis.

This may be due to its ability to block potassium ion channels which are critical to the proper function of the nervous system. However, individual responses to barium salts vary widely, with some being able to handle barium nitrate casually without problems, and others becoming ill from working with it in small quantities. For example, barium acetate was used by Marie Robards to poison her father in 1993.

Non-toxicity of Barium Sulfate

Because it is highly insoluble in water as well as stomach acids, barium sulfate can be taken orally. It is eliminated completely from the digestive tract. Unlike other heavy metals, barium does not bioaccumulate. However, inhaled dust containing barium compounds can accumulate in the lungs, causing a benign condition called baritosis.

Beryllium

Beryllium is the chemical element with the symbol Be and atomic number 4. It is a divalent element which occurs naturally only in combination with other elements in minerals. Notable gemstones which contain beryllium include beryl (aquamarine, emerald) and chrysoberyl. As a free element it is a steel-gray, strong, lightweight and brittle alkaline earth metal.

Beryllium is used primarily as a hardening agent in alloys, notably beryllium copper. In structural applications, high flexural rigidity, thermal stability, thermal conductivity and low density (1.85 times that of water) make beryllium a quality aerospace material for high-speed aircraft, missiles, space vehicles and communication satellites. Because of its low density and atomic mass, beryllium is relatively transparent to X-rays and other forms of ionizing radiation; therefore, it is the most common window material for X-ray equipment and in particle physics experiments. The high thermal conductivity of beryllium and beryllium oxide have led to their use in heat transport and heat sinking applications.

The commercial use of beryllium metal presents technical challenges due to the toxicity (especially by inhalation) of beryllium-containing dusts. Beryllium is corrosive to tissue, and can cause a chronic life-threatening allergic disease called berylliosis in some people. Because any Beryllium synthesized in stars is short-lived, it is a relatively rare element in both the Earth and the universe. The element is not known to be necessary or useful for either plant or animal life.

Characteristics

Physical Properties

Beryllium has exceptional flexural rigidity (Young's modulus 287 GPa) and a reasonably high melting point. The modulus of elasticity of beryllium is approximately 50% greater than that of steel. The combination of this modulus and a relatively low density results in an unusually fast sound conduction speed in beryllium – about 12.9 km/s at ambient conditions.

Other significant properties are high specific heat (1925 $J \cdot kg^{-1} \cdot K^{-1}$) and thermal conductivity (216 $W \cdot m^{-1} \cdot K^{-1}$), which make beryllium the metal with the best heat dissipation characteristics per unit weight. In combination with the relatively low coefficient of linear thermal expansion (11.4×10^{-6} K^{-1}), these characteristics result in a unique stability under conditions of thermal loading.

Nuclear Properties

Beryllium has a large scattering cross section for high-energy neutrons, about 6 barns for energies above ~0.01 eV. Therefore, it effectively slows the neutrons to the thermal energy range of below 0.03 eV, where the total cross section is at least an order of magnitude lower – exact value strongly depends on the purity and size of the crystallites in the material. The predominant beryllium isotope ^{9}Be also undergoes a (n,2n) neutron reaction to ^{8}Be, which then instantaneously breaks into two alpha particles; that is, beryllium is a neutron multiplier, releasing more neutrons than it absorbs. This nuclear reaction is:

$$^{9}_{4}Be + n \rightarrow 2(^{4}_{2}He) + 2n$$

As a metal, beryllium is transparent to most wavelengths of X-rays and gamma rays, making it useful for the output windows of X-ray tubes and other such apparatus. It is also a good source for the relatively-small numbers of free neutrons in the laboratory which are liberated when beryllium nuclei are struck by energetic alpha particles producing the nuclear reaction

$$^{9}_{4}Be + ^{4}_{2}He \rightarrow ^{12}_{6}C + n,$$

where $^{4}_{2}He$ is an alpha particle and $^{12}_{6}C$ is a carbon-12 nucleus.

Isotopes and nucleosynthesis

Main articles: Isotopes of beryllium and beryllium-10

Both stable and unstable isotopes of beryllium are created in stars, but these do not last long. It is believed that most of the stable beryllium in the universe was created when cosmic rays induced fission in heavier elements found in interstellar gas and dust.

Beryllium contains only one stable isotope, ^{9}Be, and therefore is a monoisotopic element. Cosmogenic ^{10}Be is produced in the atmosphere of the Earth by the cosmic ray spallation of oxygen and nitrogen. Cosmogenic ^{10}Be accumulates at the soil surface, where its relatively long half-life (1.36 million years) permits a long residence time before decaying to boron-10. Thus, ^{10}Be and its daughter products are used to examine natural soil erosion, soil formation and the development of lateritic soils, and as a proxy for measurement of the variations in solar activity and the age of ice cores.

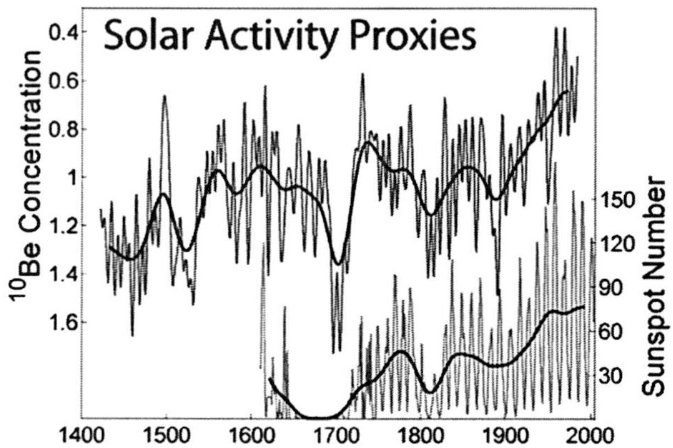

Figure 3 : Plot showing variations in solar activity, including variation in ^{10}Be concentration. Note that the beryllium scale is inverted, so increases on this scale indicate lower ^{10}Be levels

The production of ^{10}Be is inversely proportional to the solar activity, because the increased solar wind during periods of high solar magnetic activity in turn decreases the flux of galactic cosmic rays that reach the Earth. Nuclear explosions also form ^{10}Be by the reaction of fast neutrons with ^{13}C in the carbon dioxide in air. This is one of the indicators of past activity at nuclear weapon test sites.

The isotope ^{7}Be (half-life 53 days) is also cosmogenic, and shows an atmospheric abundance linked to sunspots much like ^{10}Be. ^{8}Be has a very short half-life of about 7×10^{-17} s that contributes to its significant cosmological role, as elements heavier than beryllium could not have been produced by nuclear fusion in the Big Bang. This is due to the lack of sufficient time during the Big Bang's nucleosynthesis phase to produce carbon by the fusion of ^{4}He nuclei and the very low concentrations of available beryllium-8. The British astronomer Sir Fred Hoyle first showed that the energy levels of ^{8}Be and ^{12}C allow carbon production by the so-called triple-alpha process in helium-fueled stars where more nucleosynthesis time is available, thus making creation of carbon-based life possible from the gas and dust ejected by supernovas.

The innermost electrons of beryllium may contribute to chemical bonding. Therefore, when ^{7}Be decays by electron capture, it does so by taking electrons from atomic orbitals that may participate in bonding. This makes its decay rate dependent to a measurable degree upon its electron configuration – a rare occurrence in nuclear decay.

The shortest-lived known isotope of beryllium is ^{13}Be which decays through neutron emission. It has a half-life of 2.7×10^{21} s. ^{6}Be is also very short-lived with a half-life of 5.0×10^{21} s. The exotic isotopes ^{11}Be and ^{14}Be are known to exhibit a nuclear halo. This phenomenon can be understood as the nuclei of ^{11}Be and ^{14}Be have, respectively, 1 and 4 neutrons orbiting substantially outside the classical Fermi 'waterdrop' model of the nucleus.

Occurrence

Figure 4 : Beryllium ore

Figure 5 : Emerald is a naturally occurring compound of Beryllium.

Beryllium is scarce in the universe. With a concentration of 1 part per billion (ppb) by weight, it is much less abundant than all elements preceding niobium, with the exception of boron, which has a similar abundance. Beryllium is similarly rare in the Sun with a concentration of 0.1 ppb by weight, similar to that of rhenium.

The beryllium concentration in the Earth's surface rocks is ca. 4–6 ppm by atoms. Beryllium is a major constituent of about 100 out of some 4000 known minerals, the most important of which are bertrandite ($Be_4Si_2O_7(OH)_2$), beryl ($Al_2Be_3Si_6O_{18}$), chrysoberyl (Al_2BeO_4) and phenakite (Be_2SiO_4). Precious forms of beryl are aquamarine, bixbite and emerald. In sea water, beryllium is exceedingly rare, more so than even scandium, comprising only 0.0006 ppb by weight. In stream water, however, beryllium is more abundant with 0.1 ppb by weight.

Production

Because of its high affinity for oxygen at elevated temperatures, and its ability to reduce water when its oxide film is removed, the extraction of beryllium from its compounds is a difficult process. Electrolysis of a mixture of beryllium fluoride and sodium fluoride was used to isolate beryllium during the 19th century.

The metal's high melting point makes this process more energy-consuming than corresponding processes used for the alkali metals. Early in the 20th century, the production of beryllium by the thermal decomposition of beryllium iodide was investigated following the success of a similar process for the production of zirconium, but this process proved to be uneconomical for volume production.

Pure beryllium metal did not become readily available until 1957, even though it had been used as an alloying metal to harden and toughen copper much earlier. Beryllium could be produced by reducing beryllium compounds such as beryllium chloride with metallic potassium or sodium. Currently most beryllium is produced by reducing beryllium fluoride with purified magnesium. The price on the American market for vacuum-cast beryllium ingots was about \$338 per pound (\$745 per kilogram) in 2001. The chemical equation for the key reaction is as follows:

$$BeF_2 + Mg \rightarrow MgF_2 + Be$$

The United States, China and Kazakhstan are the only three countries involved in the industrial scale extraction of beryllium. In the US, Brush Wellman Inc. is the main producer of beryllium and beryllium products. This company smelts its beryllium ore, which contains the mineral bertrandite, and which comes mostly from the company-owned Spor Mountain deposit in the State of Utah. The smelting and other refining of the beryllium is carried out at a factory 10 miles north of Delta, Utah, a location chosen for its remoteness and proximity to the Intermountain Power Project. Between 1998 and 2008, the world's production of beryllium had decreased from 343 to about 200 tonnes, of which 176 tonnes (88%) came from the United States.

Chemical Properties

Beryllium metal is located above aluminium in the electrochemical series and therefore is expected to show significant chemical activity; however, it is passivated by an oxide layer and does not react with air or water even at red heat. Once ignited, beryllium burns brilliantly forming a mixture of beryllium oxide and beryllium nitride. Beryllium dissolves readily in non-oxidizing acids, such as HCl and diluted H_2SO_4, but not in nitric acid as this forms the oxide. This behaviour is similar to that of aluminium metal. Beryllium, again similarly to aluminium, dissolves in warm alkali solutions to form the beryllate anion, $Be(OH)_4^{2-}$, and hydrogen gas.

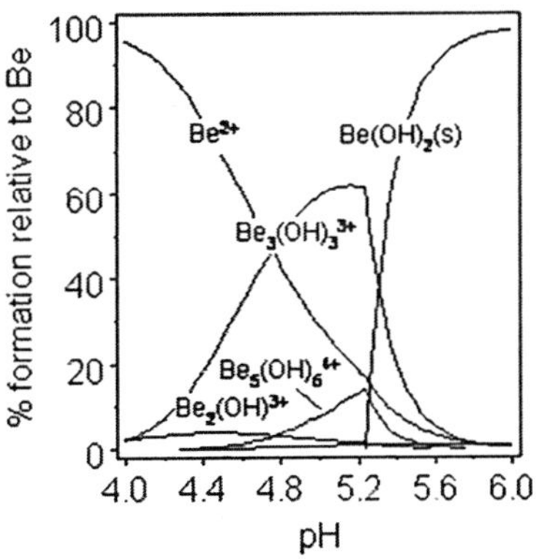

Figure 7 : Beryllium hydrolysis as a function of pH Water molecules attached to Be are omitted.

The beryllium atom has the electronic configuration [He] $2s^2$. In beryllium compounds the two electrons are lost and beryllium is in the +2 oxidation state; the only evidence of lower valence of beryllium is in the solubility of the metal in $BeCl_2$. The small atomic radius leads to significant covalent character in beryllium's bonding. Beryllium is 4-fold coordinated in virtually all of its derivatives, e.g. $[Be(H_2O)_4]^{2+}$ and tetrahaloberyllates, BeX_4^{2-}. This characteristic is employed in analytical techniques using EDTA as a ligand. EDTA preferentially forms octahedral complexes – thus absorbing other cations such as

Al^{3+} which might interfere – for example, in the solvent extraction of a complex formed between Be^{2+} and acetylacetone.

Solutions of beryllium salts, e.g. beryllium sulfate and beryllium nitrate, are acidic because of hydrolysis of the $[Be(H_2O)_4]^{2+}$ ion.

$$[Be(H_2O)_4]^{2+} + H_2O \rightleftharpoons [Be(H_2O)_3(OH)]^+ + H_3O^+$$

Other products of hydrolysis include the trimeric ion $[Be_3(OH)_3(H_2O)_6]^{3+}$. Beryllium hydroxide, $Be(OH)_2$, is insoluble even in acidic solutions with pH less than 6, that is at biological pH. It is amphoteric and dissolves in strongly alkaline solutions.

Beryllium forms binary compounds with many non-metals. Anhydrous halides are known for F, Cl, Br and I. BeF_2 has a silica-like structure with corner-shared BeF_4 tetrahedra.

$BeCl_2$ and $BeBr_2$ have chain structures with edge-shared tetrahedra. All beryllium halides have a linear monomeric molecular structure in the gas phase.

Beryllium oxide, BeO, is a white refractory solid, which has the wurtzite crystal structure and a thermal conductivity as high as in some metals. BeO is amphoteric. Salts of beryllium can be produced by treating $Be(OH)_2$ with acid. Beryllium sulfide, selenide and telluride are known, all having the zincblende structure.

Beryllium nitride, Be_3N_2 is a high-melting-point compound which is readily hydrolyzed. Beryllium azide, BeN_6 is known and beryllium phosphide, Be_3P_2 has a similar structure to Be_3N_2. Basic beryllium nitrate and basic beryllium acetate have similar tetrahedral structures with four beryllium atoms coordinated to a central oxide ion. A number of beryllium borides are known, such as Be_5B, Be_4B, Be_2B, BeB_2, BeB_6 and BeB_{12}. Beryllium carbide, Be_2C, is a refractory brick-red compound that reacts with water to give methane. No beryllium silicide has been identified.

History

Early analyses of emeralds and beryls always yielded similar elements, leading to the fallacious conclusion that both substances are aluminium silicates. René Just Haüy discovered that both crystals show strong similarities, and he asked the chemist Louis-Nicolas Vauquelin for a chemical analysis. Vauquelin was able to separate the aluminium from the beryllium by dissolving the aluminium hydroxide in an additional alkali. Vauquelin named the new element "glucinum" for the sweet taste of some of its compounds.

Friedrich Wöhler and Antoine Bussy independently isolated beryllium in 1828 by the chemical reaction of metallic potassium with beryllium chloride, as follows:

$$BeCl_2 + 2\ K \rightarrow 2\ KCl + Be$$

The potassium itself had been produced by the electrolysis of its compounds, a newly discovered process. This chemical method yielded only small grains of beryllium from which no ingot of metal could be cast or hammered.

The direct electrolysis of a molten mixture of beryllium fluoride and sodium fluoride by Paul Lebeau in 1898 resulted in the first pure samples of beryllium. It took until World War I (1914–18) before significant amounts of beryllium were produced, but large-scale production was not started until early 1930s. It saw a rapid increase during World War II, due to the rising demand for hard beryllium-copper alloys and phosphors for fluorescent lights. In the first years, most fluorescent lamps used zinc orthosilicate with varying content of beryllium to emit greenish light. Small additions of magnesium tungstate improved the blue part of the spectrum yielding acceptable white. After it was discovered that beryllium was toxic, halophosphate-based phosphors took over.

Applications

It is estimated that most beryllium is used for military applications, so information is not readily available.

Radiation Windows

Figure 8 : Beryllium target which "converts" a proton beam into a neutron beam

Figure 9: A square beryllium foil mounted in a steel case to be used as a window between a vacuum chamber and an X-ray microscope. Beryllium is highly transparent to X-rays owing to its low atomic number.

Because of its low atomic number and very low absorption for X-rays, the oldest and still one of the most important applications of beryllium is in radiation windows for X-ray tubes. Extreme demands are placed on purity and cleanliness of beryllium to avoid artifacts in the X-ray images.

Thin beryllium foils are used as radiation windows for X-ray detectors, and the extremely low absorption minimizes the heating effects caused by high intensity, low energy X-rays typical of synchrotron radiation. Vacuum-tight windows and beam-tubes for radiation experiments on synchrotrons are manufactured exclusively from beryllium.

In scientific setups for various X-ray emission studies (e.g., energy-dispersive X-ray spectroscopy) the sample holder is usually made of beryllium because its emitted X-rays have much lower energies (~100 eV) than X-rays from most studied materials.

Low atomic number also makes beryllium relatively transparent to energetic particles.

Therefore it is used to build the beam pipe around the collision region in particle physics setups, such as all four main detector experiments at the Large Hadron Collider (ALICE, ATLAS, CMS, LHCb), the Tevatron and the SLAC. The low density of beryllium allows collision products to reach the surrounding detectors without significant interaction, its stiffness allows a powerful vacuum to be produced within the pipe to minimize interaction with gases, its thermal stability allows it to function correctly at temperatures of only a few degrees above absolute zero, and its diamagnetic nature keeps it from interfering with the complex multipole magnet systems used to steer and focus the particle beams.

Mechanical Applications

Because of its stiffness, light weight and dimensional stability over a wide temperature range, beryllium metal is used for lightweight structural components in the defense and aerospace industries in high-speed aircraft, missiles, space vehicles and communication satellites. Several liquid-fuel rockets use nozzles of pure beryllium. A small number of bicycle frames were built with beryllium, at "astonishing" prices.

Beryllium is used as an alloying agent in the production of beryllium copper, which contains up to 2.5% beryllium. Beryllium-copper alloys are used in many applications because of their combination of high electrical and thermal conductivity, high strength and hardness, nonmagnetic properties, along with good corrosion and fatigue resistance. These applications include the making of spot welding electrodes, springs, non-sparking tools and electrical contacts.

Figure 10: Beryllium Copper Adjustable Wrench

The excellent elastic rigidity of beryllium has led to its extensive use in precision instrumentation, e.g. in gyroscope inertial guidance systems and in support structures for optical systems. Beryllium-copper alloys were also applied as a hardening agent in Jason pistols, which were used to strip paint from the hulls of ships.

An earlier major application of beryllium was in brakes for military aircraft because of its hardness, high melting point and exceptional heat dissipation. Environmental considerations have led to substitution by other materials.

To reduce cost, beryllium may be fabricated with large amounts of aluminium, resulting in AlBeMet (trade name). This blend is cheaper, while still retaining many desirable properties.

Mirrors

Beryllium mirrors are of particular interest. Large-area mirrors, frequently with a honeycomb support structure, are used, for example, in meteorological satellites where low weight and long-term dimensional stability are critical. Smaller beryllium mirrors are used in optical guidance systems and in fire-control systems, e.g. in the German-made Leopard 1 and Leopard 2 main battle tanks. In these systems, very rapid movement of the mirror is required which again dictates low mass and high rigidity. Usually the beryllium mirror is coated with hard electroless nickel plating which can be more easily polished to a finer optical finish than beryllium. In some applications, though, the beryllium blank is polished without any coating. This is particularly applicable to cryogenic operation where thermal expansion mismatch can cause the coating to buckle.

The James Webb Space Telescope will have 18 hexagonal beryllium sections for its mirrors. Because JWST will face a temperature of 33 K, the mirror is made of beryllium, capable of handling extreme cold better than glass. Beryllium contracts and deforms less than glass – and remains more uniform – in such temperatures. For the same reason, the optics of the Spitzer Space Telescope are entirely built of beryllium metal.

Magnetic Applications

Beryllium is non-magnetic. Therefore, tools fabricated out of beryllium are used by naval or military explosive ordnance disposal teams for work on or near naval mines, since these mines commonly have magnetic fuzes. They are also found in maintenance and construction materials near magnetic resonance imaging (MRI) machines because of the high magnetic fields generated by them. In the fields of radio communications and powerful (usually military) radars, hand tools made of beryllium are used to tune the highly magnetic klystrons, magnetrons, travelling wave tubes, etc., that are used for generating high levels of microwave power in the transmitters.

Nuclear Applications

Thin plates or foils of beryllium are sometimes used in nuclear weapon designs as the very outer layer of the plutonium pits in the primary stages of thermonuclear bombs, placed to surround the fissile material. These layers of beryllium are good "pushers" for the implosion of the plutonium-239, and they are also good neutron reflectors, just as they are in beryllium-moderated nuclear reactors.

Beryllium is also commonly used as a neutron source in laboratory experiments in which relatively few neutrons are needed (rather than having to use a nuclear reactor). For this purpose, a target of beryllium-9 is bombarded with energetic alpha particles from a radio-isotope such as polonium-210, radium-226, plutonium-239, or americium-241. In the nuclear reaction that occurs, a beryllium nucleus are transmuted into carbon-12, and one free neutron is emitted, travelling in about the same direction as the alpha particle was heading. Such neutron sources, named "urchin" neutron initiators, were used some in early atomic bombs.

Beryllium is also used at the Joint European Torus nuclear-fusion research laboratory, and it will be used in the more advanced ITER to condition the components which face the plasma. Beryllium has also been proposed as a cladding material for nuclear fuel rods, owing to its good combination of mechanical, chemical and nuclear properties. Beryllium fluoride is one of the constituent salts of the eutectic salt mixture FLiBe, which is used as a solvent, moderator and coolant in many hypothetical molten salt reactor designs.

Acoustics

Low weight and high rigidity of beryllium make it useful as a material for high-frequency speaker drivers. Because beryllium is expensive (many times more than titanium), hard to shape due to its brittleness, and toxic if mishandled, beryllium tweeters are limited to high-end home, pro audio and public address applications. More often, beryllium is alloyed with other metals, which is sometimes not disclosed for marketing purposes.

Electronic

Beryllium is a p-type dopant in III-V compound semiconductors. It is widely used in materials such as GaAs, AlGaAs, InGaAs and InAlAs grown by molecular beam epitaxy (MBE). Cross-rolled beryllium sheet is an excellent structural support for printed circuit boards in surface-mount technology. In critical electronic applications, beryllium is both a structural support and heat sink. The application also requires a coefficient of thermal expansion that is well matched to the alumina and polyimide-glass substrates. The beryllium-beryllium oxide composite "E-Materials" have been specially designed for these electronic applications and have the additional advantage that the thermal expansion coefficient can be tailored to match diverse substrate materials.

Beryllium oxide is useful for many applications that require the combined properties of an electrical insulator and an excellent heat conductor, with high strength and hardness, and a very high melting point. Beryllium oxide is frequently used as an insulator base plate in high-power transistors in radio frequency transmitters for telecommunications. Beryllium oxide is also being studied for use in increasing the thermal conductivity of uranium dioxide nuclear fuel pellets. Beryllium compounds were used in fluorescent lighting tubes, but this use was discontinued because of the disease berylliosis which developed in the workers who were making the tubes.

Complexes

Beryllium(II) readily forms complexes with strong donating ligands such as phosphine oxides and arsine oxides. There have been extensive studies of these complexes which show the stability of the O-Be bond.

Toxicity

The toxicity of beryllium depends upon the duration, intensity and frequency of exposure (features of dose), as well as the form of beryllium and the route of exposure (e.g., inhalation, dermal, ingestion). According to the International Agency for Research on Cancer (IARC), beryllium and beryllium compounds are Category 1 carcinogens; they are carcinogenic to animals (including humans). Chronic berylliosis is a pulmonary and systemic granulomatous disease caused by exposure to beryllium. Acute beryllium disease in the form of chemical pneumonitis was first reported in Europe in 1933 and in the United States in 1943. Cases of chronic berylliosis were first described in 1946 among workers in plants manufacturing fluorescent lamps in Massachusetts. Chronic berylliosis resembles sarcoidosis in many respects, and the differential diagnosis is often difficult. It killed some early workers in nuclear weapons design, such as Herbert L. Anderson.

Early researchers tasted beryllium and its various compounds for sweetness in order to verify its presence. Modern diagnostic equipment no longer necessitates this highly risky procedure and no attempt should be made to ingest this highly toxic substance. Beryllium and its compounds should be handled with great care and special precautions must be taken when carrying out any activity which could result in the release of beryllium dust (lung cancer is a possible result of prolonged exposure to beryllium laden dust). Although the use of beryllium compounds in fluorescent lighting tubes was discontinued in 1949, potential for exposure to beryllium exists in the nuclear and aerospace industries and in the refining of beryllium metal and melting of beryllium-containing alloys, the manufacturing of electronic devices, and the handling of other beryllium-containing material.

A successful test for beryllium in air and on surfaces has been recently developed and published as an international voluntary consensus standard ASTM D7202. The procedure uses dilute ammonium bifluoride for dissolution and fluorescence detection with beryllium bound to sulfonated hydroxybenzoquinoline, allowing up to 100 times more sensitive detection than the recommended limit for beryllium concentration in the workplace. Fluorescence increases with increasing beryllium concentration. The new procedure has been successfully tested on a variety of surfaces and is effective for the dissolution and ultratrace detection of refractory beryllium oxide and siliceous beryllium (ASTM D7458).

Bismuth

Bismuth is a chemical element with symbol Bi and atomic number 83. Bismuth, a trivalent poor metal, chemically resembles arsenic and antimony. Elemental bismuth may occur naturally uncombined, although its sulfide and oxide form important commercial ores. The free element is 86% as dense as lead. It is a brittle metal with a silvery white colour when newly made, but often seen in air with a pink tinge owing to the surface oxide. Bismuth metal has been known from ancient times, although until the 18th century it was often confused with lead and tin, which each have some of the metal's bulk physical properties. The etymology is uncertain but possibly comes from Arabic "bi ismid" meaning having the properties of antimony or German words *weisse masse* or *wismuth* meaning *white mass.*

Bismuth is the most naturally diamagnetic of all metals, and only mercury has a lower thermal conductivity.

Bismuth has classically been considered to be the heaviest naturally occurring stable element. Recently, however, it has been found to be very slightly radioactive: its only naturally occurring isotope bismuth-209 decays via alpha decay into thallium-205 with a half-life of more than a billion times the estimated age of the universe.

Bismuth compounds (accounting for about half the production of bismuth) are used in cosmetics, pigments, and a few pharmaceuticals. Bismuth has unusually low toxicity for a heavy metal. As the toxicity of lead has become more apparent in recent years, alloy uses for bismuth metal (presently about a third of bismuth production), as a replacement for lead, have become an increasing part of bismuth's commercial importance.

Characteristics

Physical Characteristics

Bismuth is a brittle metal with a white, silver-pink hue, often occurring in its native form with an iridescent oxide tarnish showing many colours from yellow to blue. The spiral stair stepped structure of a bismuth crystal is the result of a higher growth rate around the outside edges than on the inside edges. The variations in the thickness of the oxide layer that forms on the surface of the crystal causes different wavelengths of light to interfere upon reflection, thus displaying a rainbow of colours. When combusted with oxygen, bismuth burns with a blue flame and its oxide forms yellow fumes. Its toxicity is much lower than that of its neighbours in the periodic table such as lead, tin, tellurium, antimony, and polonium.

Although ununpentium is theoretically more diamagnetic, no other metal is verified to be more naturally diamagnetic than bismuth. It is the most diamagnetic of naturally occurring elements. (Superdiamagnetism is a different physical phenomenon.) Of any metal, it has the second lowest thermal conductivity (after mercury) and the highest Hall coefficient. It has a high electrical resistance. When deposited in sufficiently thin layers on a substrate, bismuth is a semiconductor, rather than a poor metal.

Elemental bismuth is one of very few substances of which the liquid phase is denser than its solid phase (water being the best-known example). Bismuth expands 3.32% on solidification; therefore, it was long an important component of low-melting typesetting alloys, where it compensated for the contraction of the other alloying components. Though virtually unseen in nature, high-purity bismuth can form distinctive colourful hopper crystals. Bismuth is relatively nontoxic and has a low melting point just above 271 °C, so crystals may be grown using a household stove, although the resulting crystals will tend to be lower quality than lab-grown crystals.

Chemical Characteristics

Bismuth is stable to both dry and moist air at ordinary temperatures. When red-hot it reacts with water to make bismuth(III) oxide.

$$2\ Bi + 3\ H_2O \rightarrow Bi_2O_3 + 3\ H_2$$

It reacts with large amounts of fluorine to make bismuth(V) fluoride.

$$2\ Bi + 5\ F_2 \rightarrow 2\ BiF_5$$

It reacts with small amounts of fluorine to make bismuth(III) fluoride.

$$2\ Bi + 3\ F_2 \rightarrow 2\ BiF_3$$

It also reacts with the other halogens to make bismuth(III) halides.

$$2\ Bi + 3\ Cl_2 \rightarrow 2\ BiCl_3$$

$$2\ Bi + 3\ Br_2 \rightarrow 2\ BiBr_3$$

$$2\ Bi + 3\ I_2 \rightarrow 2\ BiI_3$$

It dissolves in concentrated sulfuric acid to make bismuth(III) sulfate and sulfur dioxide.

$$6\ H_2SO_4 + 2\ Bi \rightarrow 6\ H_2O + Bi_2(SO_4)_3 + 3\ SO_2$$

It reacts with nitric acid to make bismuth(III) nitrate.

$$Bi + 6\ HNO_3 \rightarrow 3\ H_2O + 3\ NO_2 + Bi(NO_3)_3$$

It also dissolves in hydrochloric acid but only if there is oxygen present.

$$4\ Bi + 3\ O_2 + 12\ HCl \rightarrow 4\ BiCl_3 + 6\ H_2O.$$

Isotopes

The only naturally occurring isotope of bismuth, bismuth-209, was traditionally regarded as the heaviest stable isotope, but it had long been suspected to be unstable on theoretical grounds.

This was finally demonstrated in 2003 when researchers at the Institut d'Astrophysique Spatiale in Orsay, France, measured the alpha emission half-life of 209.

Bi to be 1.9×10^{19} yr, over a billion times longer than the current estimated age of the universe. Owing to its extraordinarily long half-life, for all presently known medical and industrial applications bismuth can be treated as if it is stable and non-radioactive.

The radioactivity is of academic interest because bismuth is one of few elements whose radioactivity was suspected, and indeed theoretically predicted, before being detected in the laboratory.

Bismuth has the longest known alpha decay half-life, although tellurium-128 has a double beta decay half-life of over 2.2×10^{24} yr.

Several isotopes of bismuth with short half-lives occur within the radioactive disintegration chains of actinium, radium, and thorium, and more have been synthesized experimentally.

Commercially the radioactive isotope bismuth-213 can be produced by bombarding radium with bremsstrahlung photons from a linear particle accelerator. In 1997 an antibody conjugate with Bi-213, which has a 45 minute half-life, and decays with the emission of an alpha-particle, was used to treat patients with leukemia. This isotope has also been tried in cancer treatment, e.g. in the Targeted Alpha Therapy (TAT) programme. Bismuth-213 is also found on the decay chain of Uranium-233.

New York prices (2007)

Time	*Price (USD/lb.)*
December 2000	3.85–4.15
November 2002	2.70–3.10
December 2003	2.60–2.90
June 2004	3.65–4.00
September 2005	4.20–4.60
September 2006	4.50–4.75
November 2006	6.00–6.50
December 2006	7.30–7.80
March 2007	9.25–9.75
April 2007	10.50–11.00
June 2007	18.00–19.00
November 2007	13.50–15.00

Occurrence

In the Earth's crust, bismuth is about twice as abundant as gold. It is not usually economical to mine it as a primary product. Rather, it is usually produced as a byproduct of the processing of other metal ores, especially lead, tungsten (China), tin, copper, and also silver (indirectly) or other metallic elements.

Production

The most important ores of bismuth are bismuthinite and bismite. In 2005, China was the top producer of bismuth with at least 40% of the world share followed by Mexico and Peru, reports the British Geological Survey. Native bismuth is known from Australia, Bolivia, and China.

According to the United States Geological Survey, world 2009 mine production of bismuth was 7,300 tonnes, with the major contributions from China (4,500 tonnes), Mexico (1,200 tonnes) and Peru (960 tonnes). World 2008 bismuth refinery production was 15,000 tonnes, of which China produced 78%, Mexico 8% and Belgium 5%.

The difference between world bismuth mine production and refinery production reflects bismuth's status as a byproduct metal. Bismuth travels in crude lead bullion (which can contain up to 10% bismuth) through several stages of refining, until it is removed by the Kroll-Betterton process or the Betts process. The Kroll-Betterton process uses a pyrometallurgical separation from molten lead of calcium-magnesium-bismuth drosses containing associated metals (silver, gold, zinc, some lead, copper, tellurium, and arsenic), which are removed by various fluxes and treatments to give high-purity bismuth metal (over 99% Bi). The Betts process takes cast anodes of lead bullion and electrolyzes them in a lead fluorosilicate-hydrofluorosilicic acid electrolyte to yield a pure lead cathode and an anode slime containing bismuth. Bismuth will behave similarly with another of its major metals, copper. Thus world bismuth production from refineries is a more complete and reliable statistic.

According to the Bismuth Advocate News, the price for bismuth metal from year-end 2000 to September 2005 ranged from $2.60 to $4.15 per pound, but after this period the price started rising rapidly as global bismuth demand as a lead replacement and other uses grew rapidly. New mines in Canada and Vietnam may relieve the shortages, but prices are likely to remain above their previous level for the foreseeable future. The Customer-Input price for bismuth is more oriented to the ultimate consumer; it started at US$39.40 per kilogram ($17.90 per pound) in January 2008 and reached US$35.55 per kg (US$16.15 per lb.) in September 2008.

Recycling

Whereas bismuth is most available today as a byproduct, its sustainability is more dependent on recycling. Bismuth is mostly a byproduct of lead smelting, along with silver, zinc, antimony, and other metals, and also of tungsten production, along with molybdenum and tin, and also of copper production. Recycling bismuth is difficult in many of its end uses, primarily because of scattering.

Probably the easiest to recycle would be bismuth-containing fusible alloys in the form of larger objects, then larger soldered objects. Half of the world's solder consumption is in electronics (i.e., circuit boards). As the soldered objects get smaller or contain little solder or little bismuth, the recovery gets progressively more difficult and less economic, although solder with a higher silver content will be more worthwhile recovering. Next in recycling feasibility would be sizeable catalysts with a fair bismuth content, perhaps as bismuth phosphomolybdate, and then bismuth used in galvanizing and as a free-machining metallurgical additive.

Bismuth in uses where it is dispersed most widely include stomach medicines (bismuth subsalicylate), paints (bismuth vanadate) on a dry surface, pearlescent cosmetics (bismuth oxychloride), and bismuth-containing bullets that have been fired. The bismuth scattered in these uses is unrecoverable with present technology.

Bismuth can also be available sustainably from greater efficiency of use or substitution, most likely stimulated by a rising price. It would be more difficult to find an alternative to bismuth oxychloride in cosmetics to give the pearlescent effect. However, there are many alloying formulas for solders and therefore many alternatives.

The most important sustainability fact about bismuth is its byproduct status, which can either improve sustainability (i.e., vanadium or manganese nodules) or, for bismuth from lead ore, constrain it; bismuth is constrained.

The extent that the constraint on bismuth can be ameliorated or not is going to be tested by the future of the lead storage battery, since 90% of the world market for lead is in storage batteries for gasoline or diesel-powered motor vehicles.

The life-cycle assessment of bismuth will focus on solders, one of the major uses of bismuth, and the one with the most complete information. The average primary energy use for solders is around 200 MJ per kg, with the high-bismuth solder (58% Bi) only 20% of that value, and three low-bismuth solders (2% to 5% Bi) running very close to the average.

The global warming potential averaged 10 to 14 kg carbon dioxide, with the high-bismuth solder about two-thirds of that and the low-bismuth solders about average. The acidification potential for the solders is around 0.9 to 1.1 kg sulfur dioxide equivalent, with the high-bismuth solder and one low-bismuth solder only one-tenth of the average and the other low-bismuth solders about average. There is very little life-cycle information on other bismuth alloys or compounds.

Chemical Compounds

Bismuth forms trivalent and pentavalent compounds. The trivalent compounds are more common. Many of its chemical properties are similar to those of arsenic and antimony, although they are less toxic than derivatives of those lighter elements.

Oxides and Sulfides

At elevated temperatures, the vapours of the metal combine rapidly with oxygen, forming the yellow trioxide, Bi_2O_3. On reaction with base, this oxide forms two series of oxyanions: BiO—2, which is polymeric and forms linear chains, and BiO3—3. The anion in Li_3BiO_3 is actually a cubic octameric anion, Bi_8O24—24, whereas the anion in Na_3BiO_3 is tetrameric.

The dark red bismuth(V) oxide, Bi_2O_5, is unstable, liberating O_2 gas upon heating.

Bismuth sulfide, Bi_2S_3, occurs naturally in bismuth ores. It is also produced by the combination of molten bismuth and sulfur.

Bismuthine and Bismuthides

Unlike earlier members of group 15 elements such as nitrogen, phosphorus, and arsenic, and similar to the previous group 15 element antimony, bismuth does not form a stable hydride. Bismuth hydride, bismuthine (BiH_3), is an endothermic compound that spontaneously decomposes at room temperature. It is stable only below —60°C. Bismuthides are intermetallic compounds between bismuth and other metals.

Halides

The halides of bismuth in low oxidation states have been shown to adopt unusual structures. What was originally thought to be bismuth(I) chloride, BiCl, turns out to be a complex compound consisting of Bi5+9 cations and BiCl2—5 and Bi2Cl2—8 anions. The Bi5+9 cation has a distorted tricapped trigonal prismic molecular geometry, and is also found in $Bi_{10}Hf_3Cl_{18}$, which is prepared by reducing a mixture of hafnium(IV) chloride and bismuth chloride with elemental bismuth, having the structure [Bi^+][Bi5+9][HfCl2^{-6}]^3. Other polyatomic bismuth cations are also known, such as Bi2+8, found in $Bi_8(AlCl_4)_2$. Bismuth also forms a low-valence bromide with the same structure as "BiCl". There is a *true* monoiodide, BiI, which contains chains of Bi_4I_4 units. BiI decomposes upon heating to the triiodide, BiI_3, and elemental bismuth. A monobromide of the same structure also exists.

In oxidation state +3, bismuth forms trihalides with all of the halogens: BiF_3, $BiCl_3$, $BiBr_3$, and BiI_3. All of these except BiF_3 are hydrolysed by water to form the *bismuthyl* cation, BiO+, a commonly encountered bismuth oxycation. Bismuth(III) chloride reacts with hydrogen chloride in ether solution to produce the acid $HBiCl_4$.

The oxidation state +5 is less frequently encountered. One such compound is BiF_5, a powerful oxidising and fluorinating agent. It is also a strong fluoride acceptor, reacting with xenon tetrafluoride to form the XeF+3 cation:

$$BiF_5 + XeF_4 \rightarrow XeF+3BiF—6$$

Aqueous Species

In aqueous solution, the Bi3+ion exists in various states of hydration, depending on the pH:

pH range	*Species*
<3	$Bi(H_2O)^{3+6}$
0-4	$Bi(H_2O)_5OH^{2+}$
1-5	$Bi(H_2O)_4(OH)_2+$
5-14	$Bi(H_2O)_3(OH)_3$
>11	$Bi(H_2O)_2(OH)_4-$

These mononuclear species are in equilibrium. Polynuclear species also exist, the most important of which is BiO+, which exists in hexameric form as the octahedral complex $[Bi_6O_4(OH)_4]$6+(or 6 [BiO+]·2 H_2O).

History

The name Bismuth is from ca. 1660s and is of uncertain etymology. It is one of the first 10 metals to have been discovered. Bismuth appears in the 1660s, from obsolete Ger. *Bismuth*, *Wismut*, *Wissmuth* (early 17c.); perhaps related to O.H.G. hwiz ("white"). The New Latin *bisemutum* is from German *Wismuth*, perhaps from *weibe Masse*, "white mass." The element was confused in early times with tin and lead because of its resemblance to those elements. Bismuth has been known since ancient times, so no one person is credited with its discovery. Agricola, in De Natura Fossilium states that bismuth is a distinct metal in a family of metals including tin and lead in 1546 based on observation of the metals and their physical properties. Claude François Geoffroy demonstrated in 1753 that this metal is distinct from lead and tin.

"Artificial bismuth" was commonly used in place of the actual metal. It was made by hammering tin into thin plates, and cementing them by a mixture of white tartar, saltpeter, and arsenic, stratified in a crucible over an open fire.

Bismuth was also known to the Incas and used (along with the usual copper and tin) in a special bronze alloy for knives.

Applications

Bismuth has few commercial applications, none of which are large. Taking the U.S. as an example, 1,090 tonnes of bismuth were consumed in 2008, of which 55% were chemicals (including pharmaceuticals, pigments, and cosmetics), 34% were metallurgical additives for casting and galvanizing, 7% were bismuth alloys, solders and ammunition, and the balance went for research and other uses.

In the early 1990s, researchers began to evaluate bismuth as a nontoxic replacement for lead in various applications.

Health and Cosmetics

Bismuth is an ingredient in some pharmaceuticals, although the use of some of these substances is declining. Bismuth subsalicylate is used as an antidiarrheal; it is the active ingredient in such "Pink Bismuth" preparations as Pepto-Bismol, as well as the 2004 reformulation of Kaopectate. It is also used to treat some other gastrointestinal diseases. The mechanism of action of this substance is still not well documented, although an oligodynamic effect may be involved in at least some cases. Bibrocathol is an organic bismuth-containing compound used to treat eye infections. Bismuth subgallate, the active ingredient in Devrom, is used as an internal deodorant to treat malodor from flatulence ("gas") and feces. Bismuth compounds were formerly used to treat syphilis, and today bismuth subsalicylate and bismuth subcitrate are used to treat peptic ulcers. Bismuth subnitrate and bismuth subcarbonate are also used in medicines.

Bismuth oxychloride is sometimes used in cosmetics.

Metal and Alloys

- Many bismuth alloys have low melting points and are found in specialty applications such as solders. Fire detection and suppression system safety devices widely use Bi-In-Cd–In–Sn–Pb which melts at 47 °C.
- Bismuth is also used as an alloying agent in production of malleable irons and as a thermocouple material.

- Superconductors. Bismuth is included in BSCCO (Bismuth Strontium Calcium Copper Oxide) which is a group of similar superconducting compounds discovered in 1988 among which is the highest temperature superconductor yet known with a transition temperature of 110K.
- The "RoHS" initiative for reduction of lead has broadened bismuth's use in electronics as a component of low-melting point solders, as a replacement for traditional tin-lead solders. its low toxicity will be especially important for solders to be used in food processing equipment and copper water pipes, although it can also be used in other applications including those in the automobile industry, in the EU for example.
- Bismuth has been evaluated as a replacement for lead in free-machining brasses for plumbing applications, although it doesn't equal leaded steels' performance
- It has been used to make free-machining steels for precision machining properties
- The metal has been used to replace lead as a dense material for fishing sinkers
- It has been used as a replacement for lead in shot and bullets. The Netherlands, the UK and U.S., and many other countries now prohibit the use of lead shot for the hunting of wetland birds, as many birds are prone to lead poisoning owing to mistaken ingestion of lead (instead of small stones and grit) to aid digestion, or even prohibit the use of lead for all hunting, such as in the Netherlands. Bismuth-tin alloy shot is one alternative that provides similar ballistic performance to lead. (Another less expensive but also more poorly performing alternative is "steel" shot, which is actually soft iron.) The lack of malleability does, however, make bismuth unsuitable for use in expanding hunting bullets.
- Fabrique Nationale de Herstal uses bismuth in the projectiles for its FN 303 less-lethal riot gun.

Other Uses as Compounds

- Bismuth subnitrate is a component of glazes that produces an iridescence.
- Bismuth telluride (a semiconductor) is an excellent thermoelectric material. Bi_2Te_3 diodes are used in mobile

refrigerators and CPU coolers. Also used as detectors in Infra red spectrophotometers.

- Bismuth oxide when it exists in delta form at room temperature, is a solid electrolyte for oxygen. This form normally only exists above and breaks down below a high temperature threshold, but can be electrodeposited well below this temperature in a highly alkaline solution.
- Bismuth vanadate is an opaque yellow pigment in artists' oil and acrylic paint. This pigment is generally considered to be a non-toxic lightfast substitute for lemon yellow pigments such as the cadmium sulfides and the lead/strontium/barium chromates. Unlike lead chromate+lead sulfate lemon, bismuth vanadate does not readily blacken with UV exposure.
- A catalyst for making acrylic fibres
- Ingredient in lubricating greases
- In crackling microstars (dragon's eggs) in pyrotechnics, as the oxide, subcarbonate, or subnitrate.

Toxicology and Ecotoxicology

Scientific literature concurs with the idea that bismuth and its compounds are less toxic than lead or its other periodic table neighbours (antimony, polonium) and that it is not bioaccumulative. Its biological half-life for whole-body retention is 5 days but it can remain in the kidney for years in patients treated with bismuth compounds. In the industry, it is considered as one of the least toxic heavy metals.

Bismuth poisoning exists and mostly affects the kidney and liver, and bladder. Skin and respiratory irritation can also follow exposure to respective organs. As with lead, overexposure to bismuth can result in the formation of a black deposit on the gingiva, known as a bismuth line.

Bismuth's environmental impacts are not very well known. It is considered that its environmental impact is small, due in part to the low solubility of its compounds. Limited information however means that a close eye should be kept on its impact.

Boron

Boron is the chemical element with atomic number 5 and the chemical symbol B. Boron is a metalloid. Because boron is produced entirely by cosmic ray spallation and not by stellar nucleosynthesis, it is a low-abundance element in both the solar system and the Earth's crust. However, boron is concentrated on Earth by the water-solubility of its more common naturally occurring compounds, the borate minerals. These are mined industrially as evaporate ores, such as borax and kernite.

Chemically uncombined boron is not found naturally on Earth. Industrially, very pure boron is produced with difficulty, as boron tends to form refractory materials containing small amounts of carbon or other elements. Several allotropes of boron exist: amorphous boron is a brown powder and crystalline boron is black, extremely hard (about 9.5 on Mohs' scale), and a poor conductor at room temperature. Elemental boron is used as a dopant in the semiconductor industry.

The major industrial-scale uses of boron compounds are in sodium perborate bleaches, and the borax component of fiberglass insulation. Boron polymers and ceramics play specialized roles as high-strength lightweight structural and refractory materials. Boron compounds are used in silica-based glasses and ceramics to give them resistance to thermal shock.

Boron-containing reagents are used for the synthesis of organic compounds, as intermediate in the synthesis of fine chemicals. A few boron-containing organic pharmaceuticals are used, or are in study. Natural boron is composed of two stable isotopes, one of which (boron-10) has a number of uses as a neutron-capturing agent.

In biology, borates have low toxicity in mammals (similar to table salt), but are more toxic to arthropods and are used as insecticides. Boric acid is mildly antimicrobial, and a natural boron-containing organic antibiotic is known. Boron is essential to life. Small amounts of boron compounds play a strengthening role in the cell walls of all plants, making boron necessary in soils. Experiments indicate a role for boron as an ultratrace element in animals, but the nature of its role in animal physiology is unknown.

History and Etymology

The name boron originates from the Arabic word *buraq* or the Persian word *burah*; which are names for the mineral borax.

Boron compounds were known thousands of years ago. Borax was known from the deserts of western Tibet, where it received the name of *tincal*, derived from the Sanskrit. Borax glazes were used in China from AD300, and some tincal even reached the West, where the

Persian alchemist Jobir ibn Hayyân seems to mention it in 700. Marco Polo brought some glazes back to Italy in the 13th century. Agricola, around 1600, reports the use of borax as a flux in metallurgy. In 1777, boric acid was recognized in the hot springs (soffioni) near Florence, Italy, and became known as *sal sedativum*, with mainly medical uses.

The rare mineral is called sassolite, which is found at Sasso, Italy. Sasso was the main source of European borax from 1827 to 1872, at which date American sources replaced it. Boron compounds were relatively rarely used chemicals until the late 1800s when Francis Marion Smith's Pacific Coast Borax Company first popularized these compounds and made them in volume and hence cheap

Figure 11: Sassolite

Boron was not recognized as an element until it was isolated by Sir Humphry Davy and by Joseph Louis Gay-Lussac and Louis Jacques Thénard In 1808 Davy observed that electric current sent through a solution of borates produced a brown precipitate on one of the electrodes. In his subsequent experiments he used potassium to reduce boric acid instead of electrolysis. He produced enough boron to confirm a new element and named the element *boracium*. Gay-Lussac and Thénard use iron to reduce boric acid at high temperatures. They showed by oxidizing boron with air that boric acid is an oxidation product of boron. Jöns Jakob Berzelius identified boron as an element in 1824. Pure boron was arguably first produced by the American chemist Ezekiel Weintraub in 1909.

Characteristics

Allotropes

Boron is similar to carbon in its capability to form stable covalently bonded molecular networks. Even nominally disordered (amorphous)

boron contains regular boron icosahedra which are, however, bonded randomly to each other without long-range order. Crystalline boron is a very hard, black material with a high melting point of above 2000 °C.

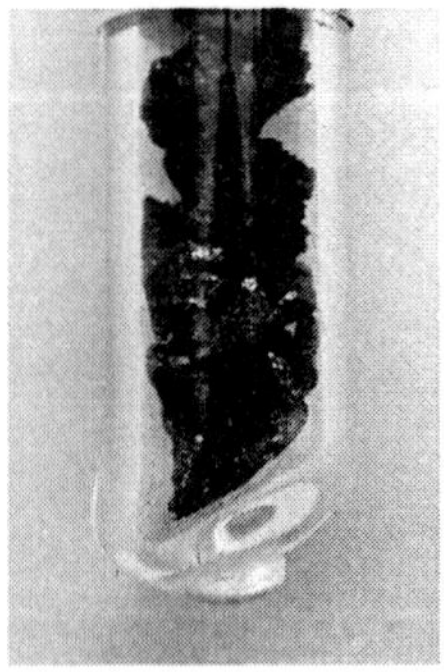

Figure 12: Boron chunks

It exists in four major polymorphs: α, β, γ and T. Whereas α, β and T phases are based on B_{12} icosahedra, the γ-phase can be described as a rocksalt-type arrangement of the icosahedra and B_2 atomic pairs. It can be produced by compressing other boron phases to 12–20 GPa and heating to 1500–1800 °C; it remains stable after releasing the temperature and pressure. The T phase is produced at similar pressures, but higher temperatures of 1800–2200 °C. As to the α and β phases, they might both coexist at ambient conditions with the β phase being more stable. Compressing boron above 160 GPa produces a boron phase with an as yet unknown structure, and this phase is a superconductor at temperatures 6–12 K.

Boron phase	***α***	***β***	***γ***	***T***
Symmetry	***Rhombohedral***	***Rhombohedral***	***Orthorhombic***	***Tetragonal***
Atoms/unit cell	12	~105	28	
Density (g/cm^3)	2.46	2.35	2.52	2.36
Vickers hardness (GPa)	42	45	50–58	
Bulk modulus (GPa)	185	224	227	
Bandgap (eV)	2	1.6	2.1	

Chemistry of the Element

Elemental boron is rare and poorly studied because the material is extremely difficult to prepare. Most studies on "boron" involve samples that contain small amounts of carbon. Chemically, boron behaves more closely to silicon than to aluminium. Crystalline boron is chemically inert and resistant to attack by boiling hydrofluoric or hydrochloric acid. When finely divided, it is attacked slowly by hot concentrated hydrogen peroxide, hot concentrated nitric acid, hot sulfuric acid or hot mixture of sulfuric and chromic acids.

The rate of oxidation of boron depends upon the crystallinity, particle size, purity and temperature. Boron does not react with air at room temperature, but at higher temperatures it burns to form boron trioxide:

$$4\ B + 3\ O_2 \rightarrow 2\ B_2O_3$$

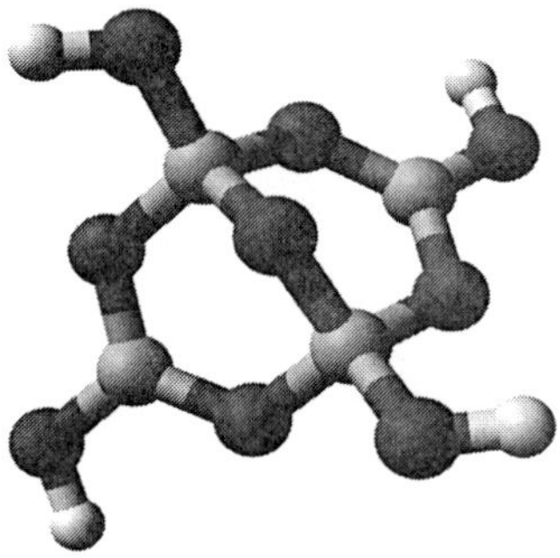

Ball-and-stick model of tetraborate anion, $[B_4O_5(OH)_4]^{2-}$, as it occurs in crystalline borax, $Na_2[B_4O_5(OH)_4]\cdot 8H_2O$. Boron atoms are pink, with bridging oxygens in red, and four hydroxyl hydrogens in white. Note two borons are trigonally bonded sp^2 with no formal charge, while the other two borons are tetrahedrally bonded sp^3, each carrying a formal charge of "1. The oxidation state of all borons is III. This mixture of boron coordination numbers and formal charges is characteristic of natural boron minerals.

Boron undergoes halogenation to give trihalides, for example:

$$2\ B + 3\ Br_2 \rightarrow 2\ BBr_3$$

These trihalides in practice are usually made from the oxides.

Chemical Compounds

In its most familiar compounds, boron has the formal oxidation state III. These include oxides, sulfides, nitrides, and halides.

The trihalides adopt a planar trigonal structure. These compounds are Lewis acids in that they readily form adducts with electron-pair donors, which are called Lewis bases. For example, fluoride (F^-) and boron trifluoride (BF_3) combined to give the tetrafluoroborate anion, BF_4^-. Boron trifluoride is used in the petrochemical industry as a catalyst. The halides react with water to form boric acid.

Boron is found in nature on Earth entirely as various oxides of B(III), often associated with other elements. The more than one hundred borates all feature boron in oxidation state +3. These mineral resemble silicates in some respect, although boron is often found not only in a tetrahedral coordination with oxygen, but also in a trigonal planar configuration. Unlike silicates, the boron minerals never feature boron with coordination number greater than four. A typical motif is exemplified by the tetraborate anions of the common mineral borax, shown at left. The formal negative charge of the tetrahedral borate centers is balanced by metal cations in the minerals, such as the sodium (Na^+) in borax.

The boron nitrides are notable for the variety of structures that they adopt. They adopt structures analogous to various allotropes of carbon, including graphite, diamond, and nanotubes. In the diamond-like structure called cubic boron nitride (tradename Borazon), boron atoms exist in the tetrahedral structure of carbons atoms in diamond, but one in every four B-N bonds can be viewed as a coordinate covalent bond, wherein two electrons are donated by the nitrogen atom which acts as the Lewis base to a bond to the Lewis acidic boron(III) centre. Cubic boron nitride, among other applications, is used as an abrasive, as it has a hardness comparable with diamond (the two substances are able to produce scratches on each other). In the BN compound analogue of graphite, hexagonal boron nitride (h-BN), the positively-charged boron and negatively-charged nitrogen atoms in each plane lie adjacent to the oppositely charged atom in the next plane. Consequently graphite and h-BN have very different properties: both are lubricants, as these planes slip past each other. However, h-BN is a relatively poor electrical and thermal conductor in the planar direction.

Organoboron Chemistry

A large number of organoboron compounds are known and many are useful in organic synthesis. Organoboron(III) compounds are usually tetrahedral or trigonal planar, for example, tetraphenylborate ($B(C_6H_5)_4^-$) vs triphenylborane ($B(C_6H_5)_3$). Many are produced from hydroboration, which employs diborane (B_2H_6).

Compounds of B(I) and B(II)

Although these are not found on Earth naturally, boron forms a variety of stable compounds with formal oxidation state less than three. As for many covalent compounds, formal oxidation states are often of little meaning in boron hydrides and metal borides. The halides also form derivatives of B(I) and B(II). BF, isoelectronic with N_2, is not isolable in condensed form, but B_2F_4 and B_4Cl_4 are well characterized.

Binary metal-boron compounds, the metal borides, feature boron in oxidation state less than III. Illustrative is magnesium diboride (MgB_2). Each boron has a formal "1 charge and magnesium is assigned a formal charge of 2+. In this material, the boron centers are trigonal planar, with an extra double bond for each boron, with the boron atoms forming sheets akin to the carbon in graphite. However, unlike the case with hexagonal boron nitride which by comparison lacks electrons in the plane of the covalent atoms, the delocalized electrons in the plane of magnesium diboride allow it to conduct electricity similar to isoelectronic graphite. In addition, in 2001 this material was found to be a high-temperature superconductor.

Certain other metal borides find specialized applications as hard materials for cutting tools.

From the structural perspective, the most distinctive chemical compounds of boron are the hydrides. Included in this series are the cluster compounds dodecaborate ($B_{12}H_{12}^{2-}$), decaborane ($B_{10}H_{14}$), and the carboranes such as $C_2B_{10}H_{12}$. Characteristically such compounds feature boron with coordination numbers greater than four.

Isotopes

Boron has two naturally occurring and stable isotopes, ^{11}B (80.1%) and ^{10}B (19.9%). The mass difference results in a wide range of ä^{11}B values, which are defined as a fractional difference between the ^{11}B and ^{10}B and traditionally expressed in parts per thousand, in natural waters ranging from "16 to +59. There are 13 known isotopes of boron, the shortest-lived isotope is ^{7}B which decays through proton emission and alpha decay. It has a half-life of 3.5×10^{-22} s. Isotopic fractionation of boron is controlled by the exchange reactions of the boron species $B(OH)_3$ and $[B(OH)_4]^-$. Boron isotopes are also fractionated during mineral crystallization, during H_2O phase changes in hydrothermal systems, and during hydrothermal alteration of rock. The latter effect results in preferential removal of the $^{10}B(OH)_4$ ion onto clays. It results in solutions enriched in $^{11}B(OH)_3$ and therefore may be responsible for the large ^{11}B enrichment in seawater relative to both oceanic crust and continental crust; this difference may act as an isotopic signature. The exotic ^{17}B exhibits a nuclear halo, i.e. its radius is appreciably larger than that predicted by the liquid drop model. The ^{10}B isotope is good at capturing thermal neutrons. Natural boron is about 20% ^{10}B and 80% ^{11}B. The nuclear industry enriches natural boron to nearly pure ^{10}B. The less-valuable by-product, depleted boron, is nearly pure ^{11}B.

Commercial Isotope Enrichment

Because of its high neutron cross-section, boron-10 is often used to control fission in nuclear reactors as a neutron-capturing substance. Several industrial-scale enrichment processes have been developed, however only the fractionated vacuum distillation of the dimethyl ether adduct of boron trifluoride (DME-BF_3) and column chromatography of borates are being used.

Enriched Boron (boron-10)

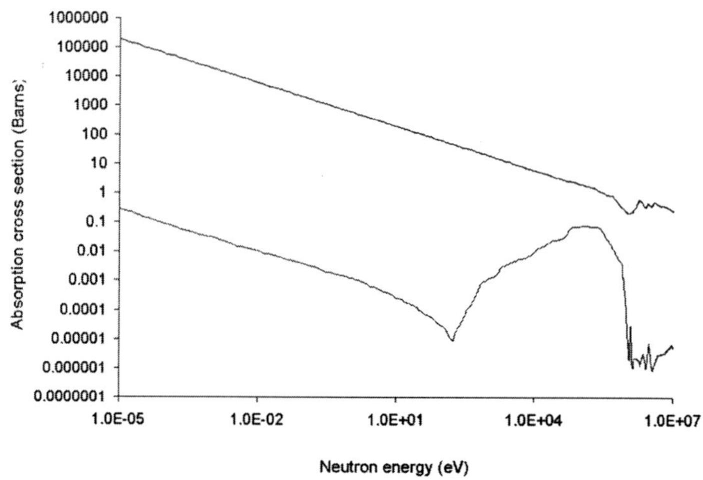

Figure 13: Neutron cross section of boron (top curve is for ^{10}B and bottom curve for ^{11}B)

Enriched boron or ^{10}B is used in both radiation shielding and in boron neutron capture therapy. In the latter, a compound containing ^{10}B is attached to a muscle near a tumour. The patient is then treated with a relatively low dose of thermal neutrons. This causes energetic and short range alpha radiation from the boron to bombard the tumour.

In nuclear reactors, ^{10}B is used for reactivity control and in emergency shutdown systems. It can serve either function in the form of borosilicate control rods or as boric acid. In pressurized water reactors, boric acid is added to the reactor coolant when the plant is shut down for refueling. It is then slowly filtered out over many months as fissile material is used up and the fuel becomes less reactive.

In future manned interplanetary spacecraft, ^{10}B has a theoretical role as structural material (as boron fibers or BN nanotube material) which would also serve a special role in the radiation shield.

One of the difficulties in dealing with cosmic rays, which are mostly high energy protons, is that some secondary radiation from interaction of cosmic rays and spacecraft materials is high energy spallation neutrons. Such neutrons can be moderated by materials high in light elements such as polyethylene, but the moderated neutrons continue to be a radiation hazard unless actively absorbed in the shielding.

Among light elements that absorb thermal neutrons, ^{6}Li and ^{10}B appear as potential spacecraft structural materials which serve both for mechanical reinforcement and radiation protection.

Depleted Boron (boron-11)

Cosmic radiation will produce secondary neutrons if it hits spacecraft structures. Those neutrons will be captured in ^{10}B, if it is present in the spacecraft's semiconductors, producing a gamma ray, an alpha particle, and a lithium ion. These resultant decay products may then irradiate nearby semiconductor 'chip' structures, causing data loss (bit flipping, or single event upset). In radiation hardened semiconductor designs, one countermeasure is to use depleted boron which is greatly enriched in ^{11}B and contains almost no ^{10}B. ^{11}B is largely immune to radiation damage. Depleted boron is a by-product of the nuclear industry.

^{11}B is also a candidate as a fuel for aneutronic fusion. When struck by a proton with energy of about 500 keV, it produces three alpha particles and 8.7 MeV of energy. Most other fusion reactions involving hydrogen and helium produce penetrating neutron radiation, which weakens reactor structures and induces long term radioactivity thereby endangering operating personnel. Whereas, the alpha particles from ^{11}B fusion can be turned directly into electric power, and all radiation stops as soon as the reactor is turned off.

NMR Spectroscopy

Both ^{10}B and ^{11}B possess nuclear spin. The nuclear spin of ^{10}B is 3 and that of ^{11}B is 3/2. These isotopes are, therefore, of use in nuclear magnetic resonance spectroscopy; and spectrometers specially adapted to detecting the boron-11 nuclei are available commercially. The ^{10}B and ^{11}B nuclei also cause splitting in the resonances of attached nuclei.

Occurrence

Figure 14: A fragment of ulexite

Figure 15 : Borax crystals

Boron is a relatively rare element in the Earth's crust, representing only 0.001%. The worldwide commercial borate deposits are estimated at 10 million tonnes.

Turkey and the United States are the world's largest producers of boron. Turkey has almost 72% of the world's boron reserves. Boron does not appear on Earth in elemental form but is found combined in borax, boric acid, colemanite, kernite, ulexite and borates. Boric acid is sometimes found in volcanic spring waters.

Ulexite is one of over a hundred borate minerals; it is a fibrous crystal where individual fibers can guide light like optical fibers.

Economically important sources of boron are rasorite (kernite) and tincal (borax ore). They are both found in the Mojave Desert of California where the Rio Tinto Borax Mine (also known as the U.S. Borax Boron Mine) Coordinates: 35°22 34.4473 N 117°402 45.4123 W near Boron, CA is California's largest open-pit mine and the largest borax mine in the world, producing nearly half the world's borates from this single site.. However, the largest borax deposits known, many still untapped, are in Central and Western Turkey including the provinces of Eskioehir, Kütahya and Balýkesir.

Production

The production of boron compounds does not involve formation of elemental boron, but exploits the convenient availability of borates.

The earliest routes to elemental boron involved reduction of boric oxide with metals such as magnesium or aluminium. However the product is almost always contaminated with metal borides. Pure boron can be prepared by reducing volatile boron halides with hydrogen at high temperatures. Ultrapure boron, for the use in semiconductor industry, is produced by the decomposition of diborane at high temperatures and then further purified with the zone melting or Czochralski processes.

Market Trend

Estimated global consumption of boron rose to a record 1.8 million tonnes of B_2O_3 in 2005, following a period of strong growth in demand from Asia, Europe and North America. Boron mining and refining capacities are considered to be adequate to meet expected levels of growth through the next decade.

The form in which boron is consumed has changed in recent years. The use of ores like colemanite has declined following concerns over arsenic content. Consumers have moved towards the use of refined borates and boric acid that have a lower pollutant content. The average cost of crystalline boron is $5/g.

Increasing demand for boric acid has led a number of producers to invest in additional capacity. Eti Mine Company of Turkey opened a new boric acid plant with the production capacity of 100,000 tonnes per year at Emet in 2003. Rio Tinto Group increased the capacity of its boron plant from 260,000 tonnes per year in 2003 to 310,000 tonnes per year by May 2005, with plans to grow this to 366,000 tonnes per year in 2006. Chinese boron producers have been unable to meet rapidly growing demand for high quality borates. This has led to imports of sodium tetraborate (borax) growing by a hundredfold between 2000 and 2005 and boric acid imports increasing by 28% per year over the same period.

The rise in global demand has been driven by high growth rates in fiberglass and borosilicate production. A rapid increase in the manufacture of reinforcement-grade fiberglass in Asia with a consequent increase in demand for borates has offset the development of boron-free reinforcement-grade fiberglass in Europe and the USA. The recent rises in energy prices may lead to greater use of insulation-grade fiberglass, with consequent growth in the boron consumption. Roskill Consulting Group forecasts that world demand for boron will grow by 3.4% per year to reach 21 million tonnes by 2010. The highest growth in demand is expected to be in Asia where demand could rise by an average 5.7% per year.

Applications

Nearly all boron ore extracted from the Earth is destined for refinement into boric acid and sodium tetraborate pentahydrate. In the United States, 70% of the boron is used for the production of glass and ceramics.

Glass and Ceramics

Borosilicate glass, which is typically 12–15% B_2O_3, 80% SiO_2, and 2% Al_2O_3, has a low coefficient of thermal expansion giving it a good resistance to thermal shock. Duran and Pyrex are two major brand names for this glass, used both in laboratory glassware and in consumer cookware and bakeware, chiefly for this resistance.

Boron filaments are high-strength, lightweight materials that are used chiefly for advanced aerospace structures as a component of composite materials, as well as limited production consumer and sporting goods such as golf clubs and fishing rods. The fibers can be produced by chemical vapour deposition of boron on a tungsten filament.

Boron fibres and sub-millimetre sized crystalline boron springs are produced by laser-assisted chemical vapour deposition. Translation of the focused laser beam allows to produce even complex helical structures. Such structures show good mechanical properties (elastic modulus 450 GPa, fracture strain 3.7%, fracture stress 17 GPa) and can be applied as reinforcement of ceramics or in micromechanical systems.

Figure 16: Borosilicate glassware. Displayed are two beakers and a test tube.

Detergent Formulations and Bleaching Agents

Borax is used in various household laundry and cleaning products, including the well-known "20 Mule Team Borax" laundry booster and "Boraxo" powdered hand soap. It is also present in some tooth bleaching formulas.

Sodium perborate serves as a source of active oxygen in many detergents, laundry detergents, cleaning products, and laundry bleaches. However, despite its name, "Borateem" laundry bleach no longer contains any boron compounds, using sodium percarbonate instead as a bleaching agent.

Insecticides

Boric acid is used as an insecticide, notably against ants, fleas, and cockroaches.

Semiconductors

Boron is a useful dopant for such semiconductors as silicon, germanium, and silicon carbide. Having one fewer valence electron than the host atom, it donates a hole resulting in p-type conductivity.

Traditional method of introducing boron into semiconductors is via its atomic diffusion at high temperatures. This process uses either solid (B_2O_3), liquid (BBr_3), or gaseous boron sources (B_2H_6 or BF_3). However, after 1970s, it was mostly replaced by ion implantation, which relies mostly on BF_3 as a boron source. Boron trichloride gas is also an important chemical in semiconductor industry, however not for doping but rather for plasma etching of metals and their oxides. Triethylborane is also injected into vapour deposition reactors as a boron source. Examples are the plasma deposition of boron-containing hard carbon films, silicon nitride-boron nitride films, and for doping of diamond film with boron.

Magnets

Boron is a component of neodymium magnets ($Nd_2Fe_{14}B$), which are the strongest type of permanent magnet. They are found in a variety of domestic and professional electromechanical and electronic devices, such as magnetic resonance imaging (MRI), various motors and actuators, computer HDDs, CD and DVD players, mobile phones, timer switches, speakers, and so on.

High-hardness and Abrasive Compounds

Several boron compounds are known for their extreme hardness and toughness.

Boron carbide and cubic boron nitride powders are widely used as abrasives. Metal borides are used for coating tools through chemical vapour deposition or physical vapour deposition. Implantation of boron ions into metals and alloys, through ion implantation or ion beam deposition, results in a spectacular increase in surface resistance and microhardness. Laser alloying has also been successfully used for the same purpose. These borides are an alternative to diamond coated tools, and their (treated) surfaces have similar properties to those of the bulk boride.

Boron Carbide

Boron carbide is a ceramic material which is obtained by decomposing B_2O_3 with carbon in the electric furnace:

$$2\ B_2O_3 + 7\ C \rightarrow B_4C + 6\ CO$$

Boron carbide's stucture is only approximately B_4C, and it shows a clear depletion of carbon from this suggested stoichiometric ratio. This is due to its very complex structure. The substance can be seen with empirical formula $B_{12}C_3$ (i.e., with B_{12} dodecahedra being a motif), but with less carbon as the suggested C_3 units are replaced with B-C chains, and there are smaller (B_6) octahedra present as well.

The repeating polymer plus semi-crystalline structure of boron carbide gives it great structural strength per weight. It is used in tank armor, bulletproof vests, and numerous other structural applications.

Boron carbide's ability to absorb neutrons without forming long-lived radionuclides (especially when doped with extra boron-10) makes the material attractive as an absorbent for neutron radiation arising in nuclear power plants.

Nuclear applications of boron carbide include shielding, control rods and shut-down pellets. Within control rods, boron carbide is often powdered, to increase its surface area.

Mechanical properties of BCN solids and ReB_2

Material	*Diamond*	*cubic-BC_2N*	*cubic-BC_5*	*cubic-BN*	*B_4C*	*ReB_2*
Vickers hardness (GPa)	115	76	71	62	38	22
Fracture toughness (MPa $m^{1/2}$)	5.3	4.5	9.5	6.8	3.5	

Other superhard Boron Compounds

- Heterodiamond (also called BCN);
- Boron nitride. This material is isoelectronic to carbon. Similar to carbon, it has both hexagonal (soft graphite-like h-BN) and cubic (hard, diamond-like c-BN) forms. h-BN is used as a high temperature component and lubricant. c-BN, also known under commercial name borazon, is a superior abrasive. Its hardness is only slightly smaller, but chemical stability is superior to that of diamond.
- Rhenium diboride can be produced at ambient pressures, but is rather expensive because of rhenium. The hardness of ReB_2 exhibits considerable anisotropy because of its hexagonal layered structure. Its value is comparable to that of tungsten carbide, silicon carbide, titanium diboride or zirconium diboride.
- $AlMgB_{14}$ + TiB_2 composites possess high hardness and wear resistance and are used in either bulk form or as coatings for components exposed to high temperatures and wear loads.

Shielding in Nuclear Reactors

Boron shielding is used as a control for nuclear reactors, taking advantage of its high cross-section for neutron capture.

Other Nonmedical Uses

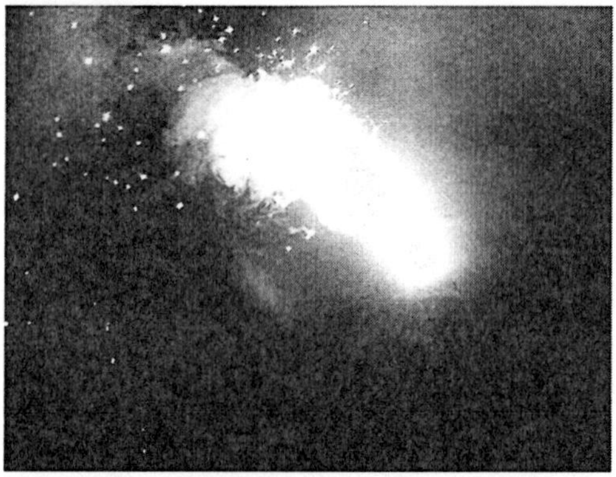

Figure 17: Boron-containing emergency flare

Figure 18: Launch of *Apollo 15* Saturn V rocket, using triethylborane ignitor

- Because of its distinctive green flame, amorphous boron is used in pyrotechnic flares. *Starch and casein-based adhesives contain sodium tetraborate decahydrate ($Na_2B_4O_7 \bullet 10\ H_2O$)
- Some anti-corrosion systems contain borax.
- Sodium borates are used as a flux for soldering silver and gold and with ammonium chloride for welding ferrous metals. They are also fire retarding additives to plastics and rubber articles.
- Boric acid (also known as orthoboric acid) H_3BO_3 is used in the production of textile fiberglass and flat panel displays and in many PVAc and PVOH based adhesives.
- Triethylborane is a substance which ignites the JP-7 fuel of the Pratt & Whitney J58 turbojet/ramjet engines powering the Lockheed SR-71 Blackbird. It was also used to ignite the F-1 Engines on the Saturn V Rocket utilized by NASA's Apollo and Skylab programs from 1967 until 1973. Triethylborane is

suitable for this because of its pyrophoric properties, especially the fact that it burns with a very high temperature. Triethylborane is an industrial initiator in radical reactions, where it is effective even at low temperatures.

Research Areas

Magnesium diboride is an important superconducting material with the transition temperature of 39 K. MgB_2 wires are produced with the powder-in-tube process and applied in superconducting magnets.

Amorphous boron is used as a melting point depressant in nickel-chromium braze alloys.

Hexagonal boron nitride forms atomically thin layers, which have been used to enhance the electron mobility in graphene devices. It also forms nanotubular structures (BNNTs), which have with high strength, high chemical stability, and high thermal conductivity, among its list of desirable properties.

Biological Role

There is a boron-containing natural antibiotic, boromycin, isolated from streptomyces. Boron is an essential plant nutrient, required primarily for maintaining the integrity of cell walls. Conversely, high soil concentrations of > 1.0 ppm can cause marginal and tip necrosis in leaves as well as poor overall growth performance.

Levels as low as 0.8 ppm can cause these same symptoms to appear in plants particularly sensitive to boron in the soil. Nearly all plants, even those somewhat tolerant of boron in the soil, will show at least some symptoms of boron toxicity when boron content in the soil is greater than 1.8 ppm. When this content exceeds 2.0 ppm, few plants will perform well and some may not survive. When boron levels in plant tissue exceed 200 ppm symptoms of boron toxicity are likely to appear.

As an ultratrace element, boron is necessary for the optimal health of rats, although it is necessary in such small amounts that ultrapurified foods and dust filtration of air is necessary to induce boron deficiency, which manifest as poor coat or hair quality. Presumably, boron is necessary to other mammals. No deficiency syndrome in humans has been described. Small amounts of boron occur widely in the diet, and the amounts needed in the diet would, by analogy with rodent studies, be very small. The exact physiological role of boron in the animal kingdom is poorly understood.

Boron occurs in all foods produced from plants. Since 1989 its nutritional value has been argued. It is thought that boron plays several biochemical roles in animals, including humans. The U.S. Department of agriculture conducted an experiment in which postmenopausal women took 3 mg of boron a day. The results showed that supplemental boron reduced excretion of calcium by 44%, and activated estrogen and vitamin D, suggesting a possible role in the suppression of osteoporosis. However, whether these effects were conventionally nutritional, or medicinal, could not be determined. The U.S. National Institutes of Health states that "Total daily boron intake in normal human diets ranges from 2.1–4.3 mg boron/day."

Analytical Quantification

For determination of boron content in food or materials the colorimetric curcumin method is used. Boron has to be transferred to boric acid or borates and on reaction with curcumin in acidic solution, a red colored boron-chelate complex, rosocyanine, is formed.

Boron Pharmaceuticals and Biologicals

Boric acid has antiseptic, antifungal, and antiviral properties and for this reasons is applied as a water clarifier in swimming pool water treatment. Mild solutions of boric acid have been used as eye antiseptics.

A number of potential boronated pharmaceuticals using boron-10, have been prepared for use in boron neutron capture therapy (BNCT).

Some boron compounds show promise in treating arthritis, though none have as yet been generally approved for the purpose.

Boron is used as an intermediate in pharmaceutical synthesis, but it appeared as an active element in its first-approved organic pharamaceutical in bortezomib, a new class of drug called proteasome inhibitors, which are active in myeloma and one form of lymphoma. The boron atom in bortezomib binds the catalytic site of the 26S proteasome with high affinity and specificity.

Health Issues

Elemental boron, boron oxide, boric acid, borates, and many organoboron compounds are non-toxic to humans and animals (approximately similar to table salt). The LD_{50} (dose at which there is 50% mortality) for animals is about 6 g per kg of body weight. Substances with LD_{50} above 2 g are considered non-toxic. The minimum lethal dose for humans has not been established, but an intake of 4 g/

day was reported without incidents, and medical dosages of 20 g of boric acid for neutron capture therapy caused no problems. Fish have survived for 30 min in a saturated boric acid solution and can survive longer in strong borax solutions. Boric acid is more toxic to insects than to mammals, and is routinely used as an insecticide.

The boranes and similar gaseous compounds are quite poisonous. As usual, it is not an element that is intrinsically poisonous, but toxicity depends on structure.

The boranes (boron hydrogen compounds) are toxic as well as highly flammable and require special care when handling. Sodium borohydride presents a fire hazard due to its reducing nature, and the liberation of hydrogen on contact with acid. Boron halides are corrosive.

Congenital endothelial dystrophy type 2, a rare form of corneal dystrophy, is linked to mutations in SLC4A11 gene that encodes a transporter reportedly regulating the intracellular concentration of boron.

Cement

In the most general sense of the word, a cement is a binder, a substance that sets and hardens independently, and can bind other materials together. The word "cement" traces to the Romans, who used the term *opus caementicium* to describe masonry resembling modern concrete that was made from crushed rock with burnt lime as binder. The volcanic ash and pulverized brick additives that were added to the burnt lime to obtain a hydraulic binder were later referred to as cementum, cimentum, cäment and cement.

Cement used in construction is characterized as hydraulic or non-hydraulic. Hydraulic cements (*e.g.,* Portland cement) harden because of hydration, chemical reactions that occur independently of the mixture's water content; they can harden even underwater or when constantly exposed to wet weather. The chemical reaction that results when the anhydrous cement powder is mixed with water produces hydrates that are not water-soluble. Non-hydraulic cements (*e.g.,* lime and gypsum plaster) must be kept dry in order to retain their strength.

The most important use of cement is the production of mortar and concrete—the bonding of natural or artificial aggregates to form a strong building material that is durable in the face of normal environmental effects.

Concrete should not be confused with cement, because the term *cement* refers to the material used to bind the aggregate materials of concrete. Concrete is a combination of a cement and aggregate.

History of the Origin of Cement

Early Uses

It is uncertain where it was first discovered that a combination of hydrated non-hydraulic lime and a pozzolan produces a hydraulic mixture, but concrete made from such mixtures was first used by the Ancient Macedonians and three centuries later on a large scale by Roman engineers. They used both natural pozzolans (trass or pumice) and artificial pozzolans (ground brick or pottery) in these concretes. Many excellent examples of structures made from these concretes are still standing, notably the huge monolithic dome of the Pantheon in Rome and the massive Baths of Caracalla. The vast system of Roman aqueducts also made extensive use of hydraulic cement.

Although any preservation of this knowledge in literary sources from the Middle Ages is unknown, medieval masons and some military engineers maintained an active tradition of using hydraulic cement in structures such as canals, fortresses, harbors, and shipbuilding facilities. The technical knowledge of making hydraulic cement was later formalized by French and British engineers in the 18th century.

Modern Cement

Modern hydraulic cements began to be developed from the start of the Industrial Revolution (around 1800), driven by three main needs:

- Hydraulic cement render (stucco) for finishing brick buildings in wet climates.
- Hydraulic mortars for masonry construction of harbour works, etc., in contact with sea water.
- Development of strong concretes.

In Britain particularly, good quality building stone became ever more expensive during a period of rapid growth, and it became a common practice to construct prestige buildings from the new industrial bricks, and to finish them with a stucco to imitate stone. Hydraulic limes were favoured for this, but the need for a fast set time encouraged the development of new cements. Most famous was Parker's "Roman cement". This was developed by James Parker in the 1780s, and finally patented in 1796. It was, in fact, nothing like any material used by the Romans, but was a "Natural cement" made by burning septaria – nodules that are found in certain clay deposits, and that contain both clay minerals and calcium carbonate. The burnt nodules were ground to a fine powder. This product, made into a mortar with sand, set in 5–15 minutes. The success of "Roman Cement" led other manufacturers to develop rival products by burning artificial mixtures of clay and chalk.

John Smeaton made an important contribution to the development of cements when he was planning the construction of the third Eddystone Lighthouse (1755–9) in the English Channel. He needed a hydraulic mortar that would set and develop some strength in the twelve hour period between successive high tides. He performed an exhaustive market research on the available hydraulic limes, visiting their production sites, and noted that the "hydraulicity" of the lime was directly related to the clay content of the limestone from which it was made. Smeaton was a civil engineer by profession, and took the idea no further. Apparently unaware of Smeaton's work, the same principle was identified by Louis Vicat in the first decade of the nineteenth century. Vicat went on to devise a method of combining chalk and clay into an intimate mixture, and, burning this, produced an "artificial cement" in 1817. James Frost, working in Britain, produced what he called "British cement" in a similar manner around the same time, but did not obtain a patent until 1822. In 1824, Joseph Aspdin patented a similar material, which he called Portland cement, because the render made from it was in color similar to the prestigious Portland stone.

Setting time and "early strength" are important characteristics of cements. Hydraulic limes, "natural" cements, and "artificial" cements all rely upon their belite content for strength development. Belite develops strength slowly. Because they were burned at temperatures below 1250 °C, they contained no alite, which is responsible for early strength in modern cements. The first cement to consistently contain alite was made by Joseph Aspdin's son William in the early 1840s. This was what we call today "modern" Portland cement. Because of the air of mystery with which William Aspdin surrounded his product, others (*e.g.*, Vicat and I.C. Johnson) have claimed precedence in this invention, but recent analysis of both his concrete and raw cement have shown that William Aspdin's product made at Northfleet, Kent was a true alite-based cement. However, Aspdin's methods were "rule-of-thumb": Vicat is responsible for establishing the chemical basis of these cements, and Johnson established the importance of sintering the mix in the kiln.

William Aspdin's innovation was counterintuitive for manufacturers of "artificial cements", because they required more lime in the mix (a problem for his father), a much higher kiln temperature (and therefore more fuel), and the resulting clinker was very hard and rapidly wore down the millstones, which were the only available grinding technology of the time.

Manufacturing costs were therefore considerably higher, but the product set reasonably slowly and developed strength quickly, thus opening up a market for use in concrete. The use of concrete in construction grew rapidly from 1850 onwards, and was soon the dominant use for cements. Thus Portland cement began its predominant role.

In the US the first large scale use of cement was Rosendale cement a natural cement mined from a massive deposit of a large dolostone rock deposit discovered in the early 19th century near Rosendale, New York.

Rosendale cement was extremely popular for the foundation of buildings (*e.g.*, Statue of Liberty, Capitol Building, Brooklyn Bridge) and lining water pipes. But its long curing time of at least a month made it unpopular after World War One in the construction of highways and bridges and many states and construction firms turned to the use of Portland cement. Because of the switch to Portland cement, by the end of the 1920s of the 15 Rosendale cement companies, only one had survived.

But in the early 1930s it was soon discovered that Portland cement while it had a faster setting time was not as durable, especially for highways, to the point that some states stopped building highways and roads with cement. An engineer, Bertrain H. Wait, whose company had worked on the construction of the New York Cities Catskill Aqueduct, and was impressed with the durability of Rosendale cement, came up with a blend of both Rosendale and synthetic cements which has the good attributes of both: it was highly durable and had a much faster setting time. Mr. Wait convinced the New York Commissioner of Highways to construct an experimental section highway near New Paltz, New York, of one sack of Rosendale to six sacks of synthetic cement, and it was proved a success and for decades hence the Rosendale-synthetic cement blend became common use in highway and bridge construction.

Types of Modern Cement

Portland Cement

Cement is made by heating limestone (calcium carbonate) with small quantities of other materials (such as clay) to 1450 °C in a kiln, in a process known as calcination, whereby a molecule of carbon dioxide is liberated from the calcium carbonate to form calcium oxide, or quicklime, which is then blended with the other materials that have been included in the mix. The resulting hard substance, called 'clinker', is then ground with a small amount of gypsum into a powder to make 'Ordinary Portland Cement', the most commonly used type of cement (often referred to as OPC).

Portland cement is a basic ingredient of concrete, mortar and most non-speciality grout. The most common use for Portland cement is in the production of concrete. Concrete is a composite material consisting of aggregate (gravel and sand), cement, and water. As a construction material, concrete can be cast in almost any shape desired, and once hardened, can become a structural (load bearing) element. Portland cement may be grey or white.

Portland Cement Blends

Portland cement blends are often available as inter-ground mixtures from cement manufacturers, but similar formulations are often also mixed from the ground components at the concrete mixing plant.

Portland blastfurnace cement contains up to 70 % ground granulated blast furnace slag, with the rest Portland clinker and a little gypsum. All compositions produce high ultimate strength, but as slag content is increased, early strength is reduced, while sulfate resistance increases and heat evolution diminishes. Used as an economic alternative to Portland sulfate-resisting and low-heat cements.

Portland flyash cement contains up to 30 % fly ash. The fly ash is pozzolanic, so that ultimate strength is maintained. Because fly ash addition allows a lower concrete water content, early strength can also be maintained. Where good quality cheap fly ash is available, this can be an economic alternative to ordinary Portland cement.

Portland pozzolan cement includes fly ash cement, since fly ash is a pozzolan, but also includes cements made from other natural or artificial pozzolans. In countries where volcanic ashes are available (e.g. Italy, Chile, Mexico, the Philippines) these cements are often the most common form in use. Portland silica fume cement. Addition of silica fume can yield exceptionally high strengths, and cements containing 5–20 % silica fume are occasionally produced. However, silica fume is more usually added to Portland cement at the concrete mixer.

Masonry cements are used for preparing bricklaying mortars and stuccos, and must not be used in concrete. They are usually complex proprietary formulations containing Portland clinker and a number of other ingredients that may include limestone, hydrated lime, air entrainers, retarders, waterproofers and colouring agents. They are formulated to yield workable mortars that allow rapid and consistent masonry work. Subtle variations of Masonry cement in the US are Plastic Cements and Stucco Cements. These are designed to produce controlled bond with masonry blocks.

Expansive cements contain, in addition to Portland clinker, expansive clinkers (usually sulfoaluminate clinkers), and are designed to offset the effects of drying shrinkage that is normally encountered with hydraulic cements. This allows large floor slabs (up to 60 m square) to be prepared without contraction joints. White blended cements may be made using white clinker and white supplementary materials such as high-purity metakaolin. Colored cements are used for decorative purposes. In some standards, the addition of pigments to produce "colored Portland cement" is allowed. In other standards (e.g. ASTM), pigments are not allowed constituents of Portland cement, and colored cements are sold as "blended hydraulic cements".

Very finely ground cements are made from mixtures of cement with sand or with slag or other pozzolan type minerals that are extremely finely ground together. Such cements can have the same physical characteristics as normal cement but with 50% less cement particularly due to their increased surface area for the chemical reaction. Even with intensive grinding they can use up to 50% less energy to fabricate than ordinary Portland cements.

Non-Portland Hydraulic Cements

Pozzolan-lime cements. Mixtures of ground pozzolan and lime are the cements used by the Romans, and can be found in Roman structures still standing (e.g. the Pantheon in Rome). They develop strength slowly, but their ultimate strength can be very high. The hydration products that produce strength are essentially the same as those produced by Portland cement.

Slag-lime cements. Ground granulated blast furnace slag is not hydraulic on its own, but is "activated" by addition of alkalis, most economically using lime. They are similar to pozzolan lime cements in their properties. Only granulated slag (i.e. water-quenched, glassy slag) is effective as a cement component.

Supersulfated cements. These contain about 80% ground granulated blast furnace slag, 15 % gypsum or anhydrite and a little Portland clinker or lime as an activator. They produce strength by formation of ettringite, with strength growth similar to a slow Portland cement. They exhibit good resistance to aggressive agents, including sulfate.

Calcium aluminate cements are hydraulic cements made primarily from limestone and bauxite. The active ingredients are monocalcium aluminate $CaAl_2O_4$ ($CaO \cdot Al_2O_3$ or CA in Cement chemist notation, CCN) and mayenite $Ca_{12}Al_{14}O_{33}$ ($12\ CaO \cdot 7\ Al_2O_3$, or $C_{12}A_7$ in CCN). Strength forms by hydration to calcium aluminate hydrates. They are well-adapted for use in refractory (high-temperature resistant) concretes, e.g. for furnace linings.

Calcium sulfoaluminate cements are made from clinkers that include ye'elimite ($Ca_4(AlO_2)_6SO_4$ or $C_4A_3\bar{S}$ in Cement chemist's notation) as a primary phase. They are used in expansive cements, in ultra-high early strength cements, and in "low-energy" cements. Hydration produces ettringite, and specialized physical properties (such as expansion or rapid reaction) are obtained by adjustment of the availability of calcium and sulfate ions. Their use as a low-energy alternative to Portland cement has been pioneered in China, where several million tonnes per year are produced. Energy requirements are lower because of the lower kiln temperatures required for reaction, and the lower amount of limestone (which must be endothermically decarbonated) in the mix. In addition, the lower limestone content and lower fuel consumption leads to a CO_2 emission around half that associated with Portland clinker. However, SO_2 emissions are usually significantly higher.

"Natural" cements correspond to certain cements of the pre-Portland era, produced by burning argillaceous limestones at moderate temperatures. The level of clay components in the limestone (around 30–35 %) is such that large amounts of belite (the low-early strength, high-late strength mineral in Portland cement) are formed without the formation of excessive amounts of free lime. As with any natural material, such cements have highly variable properties.

Geopolymer cements are made from mixtures of water-soluble alkali metal silicates and aluminosilicate mineral powders such as fly ash and metakaolin.

The Setting of Cement

Cement sets when mixed with water by way of a complex series of hydration chemical reactions still only partly understood. The different constituents slowly hydrate and crystallise while the interlocking of their crystals gives to cement its strength. After the initial setting, immersion in warm water will speed up setting. In Portland cement, gypsum is added as a compound preventing cement flash setting. The time it takes for cement to set varies; and can take anywhere from twenty minutes for initial set, to twenty-four hours, or more, for final set.

Safety Issues

Bags of cement routinely have health and safety warnings printed on them because not only is cement highly alkaline, but the setting process is exothermic. As a result, wet cement is strongly caustic and can easily cause severe skin burns if not promptly washed off with water. Similarly, dry cement powder in contact with mucous membranes can cause severe eye or respiratory irritation. Cement users should wear protective clothing.

Cement Industry in the World

In 2010 the world production of hydraulic cement was 3,300 million tonnes. The top three producers were China with 1,800, India with 220 and USA with 63.5 million tonnes for a combined total of over half the world total by the world's three most populated states.

For the world capacity to produce cement in 2010 the situation was similar with the top three states (China, India and USA) accounting for just under half the world total capacity.

China

"For the past 18 years, China consistently has produced more cement than any other country in the world. [...] (However,) China's cement export peaked in 1994 with 11 million tonnes shipped out and has been in steady decline ever since. Only 5.18 million tonnes were exported out of China in 2002. Offered at $34 a ton, Chinese cement is pricing itself out of the market as Thailand is asking as little as $20 for the same quality."

In 2006 it was estimated that China manufactured 1.235 billion tonnes of cement, which was 44% of the world total cement production.

"Demand for cement in China is expected to advance 5.4% annually and exceed 1 billion tonnes in 2008, driven by slowing but healthy growth in construction expenditures. Cement consumed in China will amount to 44% of global demand, and China will remain the world's largest national consumer of cement by a large margin."

In 2010, 3.3 billion tonnes of cement was consumed globally. Of this, China accounted for 1.8 billion tonnes.

Africa

Environmental Impacts

Cement manufacture causes environmental impacts at all stages of the process. These include emissions of airborne pollution in the form of dust, gases, noise and vibration when operating machinery and during blasting in quarries, and damage to countryside from quarrying. Equipment to reduce dust emissions during quarrying and manufacture of cement is widely used, and equipment to trap and separate exhaust gases are coming into increased use. Environmental protection also includes the re-integration of quarries into the countryside after they have been closed down by returning them to nature or re-cultivating them.

CO_2 Emissions

Cement manufacturing releases CO_2 in the atmosphere both directly when calcium carbonate is heated, producing lime and carbon dioxide, and also indirectly through the use of energy if its production involves the emission of CO_2. The cement industry is the second largest CO_2 emitting industry behind power generation. The cement industry produces about 5% of global man-made CO_2 emissions, of which 50% is from the chemical process, and 40% from burning fuel. The amount of CO_2 emitted by the cement industry is nearly 900 kg of CO_2 for every 1000 kg of cement produced. The high proportion of carbon dioxide produced in the chemical reaction leads to large decrease in mass in the conversion from limestone to cement. So, to reduce the transport of heavier raw materials and to minimise the associated costs, it is more economical for cement plants to be closer to the limestone quarries rather than to the consumer centres.

In certain applications, lime mortar, reabsorbs the same amount of CO_2 as was released in its manufacture, and has a lower energy requirement in production than mainstream cement. Newly developed cement types from Novacem and Eco-cement can absorb carbon dioxide from ambient air during hardening. Use of the Kalina cycle during production can also increase energy efficiency.

Heavy Metal Emissions in the Air

In some circumstances, mainly depending on the origin and the composition of the raw materials used, the high-temperature calcination process of limestone and clay minerals can release in the atmosphere gases and dust rich in volatile heavy metals, a.o, thallium, cadmium and mercury are the most toxic. Heavy metals (Tl, Cd, Hg,...) are often found as trace elements in common metal sulfides (pyrite (FeS_2), zinc blende (ZnS), galena (PbS),...) present as secondary minerals in most of the raw materials. Environmental regulations exist in many countries to limit these emissions. As of 2011 in the United States, cement kilns are "legally allowed to pump more toxins into the air than are hazardous-waste incinerators."

Heavy Metals Present in the Clinker

The presence of heavy metals in the clinker arises both from the natural raw materials and from the use of recycled by-products or alternative fuels. The high pH prevailing in the cement porewater ($12.5 < pH < 13.5$) limits the mobility of many heavy metals by decreasing their solubility and increasing their sorption onto the cement mineral phases. Nickel, zinc and lead are commonly found in cement in non-negligible concentrations.

Use of Alternative Fuels and by-products Materials

A cement plant consumes 3 to 6 GJ of fuel per tonne of clinker produced, depending on the raw materials and the process used. Most cement kilns today use coal and petroleum coke as primary fuels, and to a lesser extent natural gas and fuel oil. Selected waste and by-products with recoverable calorific value can be used as fuels in a cement kiln, replacing a portion of conventional fossil fuels, like coal, if they meet strict specifications. Selected waste and by-products containing useful minerals such as calcium, silica, alumina, and iron can be used as raw materials in the kiln, replacing raw materials such as clay, shale, and limestone. Because some materials have both useful mineral content and recoverable calorific value, the distinction between alternative fuels and raw materials is not always clear. For example, sewage sludge has a low but significant calorific value, and burns to give ash containing minerals useful in the clinker matrix.

Bibliography

Albritton, C. C. : *The Fabric of Geology: Reading*, Mass., Addison-Wesley Publishing Co., 2013.

Armistead, George: *Safety in Petroleum Refining and Related Industries*, New York: John G. Simmons & Co., 2000.

Arora, G D : *Liquid Crystals and Polymers*, Sarup and Sons, Delhi, 2005.

Arun Kumar Biswas: *Minerals and Metals in Pre-Modern India*, D K Printworld, Delhi, 2001.

Balakrishan, A. : *Biofuel Crops and Renewable Energy*, Sublime Publications, Delhi, 2011.

Barker, Daniel S.: *Igneous Rocks*, Prentice-Hall, Delhi, 2013.

Barnes, Thomas G. : *Origin and Destiny of the Earth's Magnetic Field*, Creation Research Society, London, 2013.

Belknap, D.F.: *Sea-level Rise on the Maine Islands: Proceedings*, Sixth Island Institute Conference, Hurricane Island, Maine, Island Institute, Rockland, ME, 2008.

Berlin, G.L.: *Earthquakes and the Urban Environment*, CRC Press, Boca Raton, FL, 2010.

Bhakar S.R. : *Ground Water Hydrology : Theory and Practice*, Agrotech, Delhi, 2009.

Biggar, John: *The Andes: A Guide for Climbers*, Andes Publishing, Scotland, 2005.

Binay Kumar and R.P. Tandon: *Advances in Technologically Important Crystals*, Macmillan Publishers India, Delhi, 2007.

Biyani, A.K. : *Dimensions of Himalayan Geology*, Satish Serial Pub, Delhi, 2007.

Bogli, Alfred: *Karst Hydrology and Physical Speleology*, Springer-Verlag, New York, 2010.

Bromley, Simon: *American Hegemony and World Oil: The Industry, the State System, and the World Economy*. University Park, Pa., 2011.

Chowdhary, L.R. : *Petroleum Geology of the Cambay Basin, Gujarat, India*, Indian Petroleum, Delhi, 2004.

Cotton, C.A.: *Volcanoes as Landscape Forms*, Hafner Publishing. New York, 2015.

DeLaisi, Francis: *Oil, Its Influence on Politics*. Labor Publishing,2003.

Doyle, F. L.: *Karst Hydrogeology,* The University of Alabama in Huntsville Press, Huntsville, Alabama, 2007.

Dreybrodt, Wolfgang: *Processes in Karst Systems Physics, Chemistry and Geology,* Springer-Verlag, New York, 2014.

Dunai, T.J.: *Cosmogenic Nucleides,* Cambridge University Press, U.K., 2010.

Eaton, Rev. S.J.M.: *Petroleum, a History of the Oil Region*, J.P. Skelly & Co., Philadelphia, 2006.

Edward W.: *United States Oil Policy and Diplomacy: A Twentieth-Century Overview.* West-port, Conn.: Greenwood Press, 2013.

Edwards, Denis : *Earth Revealing Earth Healing: Ecology and Christian Theology,* Collegeville: Liturgical Press, 2001.

Edwards, Edward : *Pattern and Design with Dynamic Symmetry,* NY: Dover 2007.

Fischer, Louis: *Oil Imperialism: The International Struggle for Petroleum.* New York: International Publishers, 2006.

Folk, R.L.: *The Petrology of Sedimentary Rocks*, Hemphill Publishing Co., Austin, 2001.

Fritz Helmet: *Diffraction in Crystals*, Ivy Publishing House, Delhi, 2011.

Garverick, Linda: *Corrosion in the Petrochemical Industry.* Materials Park, Ohio: ASM International, 2000.

Ghosh, N.C. and K.D. Sharma: *Groundwater Modelling and Management*, Capital Pub, Delhi, 2006.

Glenn T. Trewartha : *Elements of Geography*, New York, McGraw Hill, 2001.

Gontz, A.M.: *Evolution of Seabed Pockmarks in Penobscot Bay, Maine*, University of Maine, Orono, 2011.

Gubbins, David: *Seismology and Plate Tectonics*, Cambridge University Press, North America, 2010.

Günter Simader: *Fuel Cells and Their Applications*. New York: VCH, 1996.

Gurjar, Ram Kumar : *Geography of Water Resources*, Rawat, Delhi, 2008.

Gurugnanam, B. : *Essentials of Hydrogeology*, New India Pub, Delhi, 2009.

Hall, Anthony: *Igneous Petrology,* Longman, New York, 2013.

Hayward, J.A.: *Hydrology and Stream Sediment from Torlesse Stream Catchment*, Oxford University, UK, 2012.

Holmgren, D. A., Moody, J. D. : *The Structural Setting for Giant oil and Gas Fields*, Tokyo, 2015.

Inkpen, Robert: *Science, Philosophy and Physical Geography*, Routledge, London.

Jakucs, Laszlo: *Morphogenetics of Karst Regions: Variants of Karst Evolution,* Halstead Press, John Wiley and Sons, New York, 2016.

John C. Grover.: *Volcanic Eruptions and Great Earthquakes*, Copyright Publishing Company, Australia, 2016.

Kumar Yadav, Surendra : *Soil Ecology : Remote Sensing and Geographic Information System in Management,* APH, Delhi, 2007.

Lovesun, V.J. ; P.K. Sen and Amalendu Sinha: *Exploration, Exploitation, Enrichment and Environment of Coastal Placer Minerals*, Macmillan Publishers India, Delhi, 2007.

Mahajan Gautam : *Ground Water : Surveys and Investigation*, APH, Delhi, 2009.

Majumdar P.K. Basu: *Law of Mines and Minerals,* Universal Law, Delhi, 2011.

Malhotra, Raman : *Mines and Minerals Laws*, Global India, Delhi, 2011.

Mallik, T K : *Marine Geology : A Scenario Around Indian Coasts*, New Academic Pub, Delhi, 2008.

Mandelbrot, Benoit : *The Fractal Geometry of Nature,* San Francisco: Freeman, 2012.

Manthorp, P.A.: *Estuarine Circulation and Sediment Transport in the Saco River estuary*, Boston University, Boston, MA, 2015.

Mary Lambert: *Crystal Energy*, Cico Books, Delhi, 2010.

McLaurin, John J.: *Sketches in Crude-oil*, J. Horace McFarland Company, Harrisburg, 2016.

Michael Mugerwa: *Fuel Cell Systems*. New York: Plenum Press, 2013.

Middlemost, Eric A. K.: *Magmas and Magmatic Rocks: An Introduction to Igneous Petrology*, Longman, New York, 2015.

Miller, J.R.: *Process Geomorphology*, Wm.C. Brown Publishers, Dubuque, Iowa, 2015.

Poonam Johri: *Minerals and Water*, Sonali, Delhi, 2005.

Postnikov, M. : *Lectures in Geometry Semester I Analytic Geometry*, Moscow: MIR publishers, 2012.

Pramod O. Alexander: *A Handbook of Minerals, Crystals, Rocks and Ores,* New India Pub Agency, Delhi, 2009.

Ramasamy, S. : *Remote Sensing in Water Resources,* Rawat, 2005.

Rathi, Rakesh : *Petroleum Refining Processes*, SBS Pub, Delhi, 2007.

Raymond M. : *Laser Remote Sensing: Fundamentals and Applications*, Malabar, Fla., Krieger Pub. 2012.

Reiter, L.: *Earthquake Hazard Analysis: Issues and Insights*, Columbia University Press, New York, 2010.

Richards, John A. : *Remote Sensing Digital Image Analysis: An Introduction*, Berlin, New York, 2013.

Samuel Delvin: *Vitamins, Minerals and Hormones*, Ivy Pub, Delhi, 2003.

Santhi, R and K M Sellamuthu: *Fundamentals of Forest Soils*, Satish Serial Pub, Delhi, 2008.

Sawant, P. T. : *Engineering and General Geology*, New India Publishing Agency, Delhi, 2011.

Selby, Michael John: *Earth's Changing Surface: an Introduction to Geomorphology*, Clarendon Press, Oxford, 2015.

Shearer, Peter M.: *Introduction to Seismology*, Cambridge University Press, North America, 2016.

Sidney Russell Cooke: *The Oil Trusts and Anglo-American Relations*. Macmillan, 2014.

Speight, James G.: *The Chemistry and Terminology of Petroleum*. New York: Marcel Dekker, 2010.

Steven, A. : *Grand Canyon: Monument to Catastrophe,* Institute for Creation Research, London,2014.

Summerfield, M.A.: *Global Geomorphology*, Pearson Education Ltd, Delhi, 2011.

Talukdar, A. K. : *Catalysis in Petroleum and Petrochemical Industries*, Alpha Science International, Delhi, 2005.

Tiwari, S.K. : *Ore Geology, Economic Minerals And Mineral Economics*, Atlantic Publishers, Delhi, 2010.

Tracy, Robert J.; Owens, Brent: *Petrology: Igneous, Sedimentary, and Metamorphic,* New York: W. H. Freeman, 2005.

Vijayalakshmi, G S : *Fuels and Biofuels*, Agrobios, Delhi, 2007.

Wahab, M. A. : *Essentials of Crystallography*, Narosa, Delhi, 2009.

Wahrhaftig, Clyde: *A Streetcar to Subduction: and Other Plate Tectonic Trips by Public Transportation in San Francisco*, American Geophysical Union, Washington, D.C., 2014.

Wainwright, John; Mulligan, M.: *Environmental Modelling: Finding Simplicity in Complexity*, John Wiley and Sons Ltd, London, 2003.

White, William B.: *Geomorphology and Hydrology of Karst Terrains,* Oxford University Press, New York, 2015.

Whitmore, G : *Chemical Crystallography And Lipid Crystals*, Sarup Book, Delhi, 2009.

Williams, Paul: *Karst Geomorphology and Hydrology,* Unwin Hyman, Winchester, Massachusetts, 2014.

Young, George J.: *Fuel Cells,* New York: Reinhold Publishing Corp., 2011.

Index

A

Abernathyite, 64.
Abhurite, 29, 64.
Abswurmbachite, 64.
Acuminite, 28, 67.
Adamite, 67.
Adamsite, 45, 68.
Admontite, 48, 68.
Aeschynite, 5, 34, 71.
Agardite, 73.
Aluminium Alloys, 162, 173, 174.
Aluminium Compounds, 162, 168, 170, 173, 178, 179.
Amber Perfumery, 98.
Arsenite Minerals, 38.
Asbestos Construction, 146.
Associated Fibres, 129.

B

Barium Compounds, 203, 205, 207, 208.
Barium Sulfate, 10, 203, 204, 206, 207, 208.
Binary Compounds, 214.
Bleaching Agents, 243.
Borate Minerals, 10, 46, 232, 240.
Boron Carbide, 244, 245.

C

Calcite Alabaster, 82, 83.
Carbonate Class, 10, 43.
Carbonate Minerals, 10, 40, 160.
Cement Industry, 256.
Chemical Formulae, 183.
Chemical Properties of Minerals, 9.
Commercial Isotope Enrichment, 238.
Contact Metamorphism, 72, 77, 183.
Crystal Chemistry, 106.
Crystal Structure, 2, 3, 4, 7, 50, 56, 66, 72, 77, 88, 131, 142, 154, 160, 161, 166, 186, 197, 202, 214, 218, 254, 258.

D

Distribution, 92, 112, 171.

E

Electric Power, 166, 239.
Element Class, 11.
Environmental Considerations, 217.

F

Flame Retardant, 101.
Fluorite Structure, 26.

G

Gemstone Quality, 117, 124.
Gypsum, 7, 8, 10, 50, 53, 81, 82, 83, 84, 85, 191, 192, 249, 253, 255, 256.
Gypsum Alabaster, 82, 83.

H

Halide Class, 10, 11.
Halide Minerals, 11, 26, 27.
Hard Materials, 237.
Health Problems, 136.

Heavy Metal Emissions, 258.
Hydrated Carbonates, 42, 249.

I

Industrial Processes, 98, 172.
Industrial Revolution, 104, 250.
International Amber Association, 95.
Intrusive Igneous Rocks, 66.

L

Lead Poisoning, 230.
Liquid Crystals, 3.

M

Magnetic Applications, 218.
Mechanical Applications, 217.
Metal Sulfides, 17, 18, 19, 20.
Metamorphic Processes, 184.
Metamorphic Rock, 121.
Mineral Composition of Rocks, 6.
Mining Operations, 116.

N

Natural Occurrence, 163, 201.
Nuclear Applications, 218, 245.
Nuclear Industry, 238, 239.
Nuclear Reactors, 218, 238, 245.

O

Ore Deposits, 3, 161, 181, 192.
Organic Class, 2, 12.
Oxide Class, 11, 30.
Oxide Minerals, 30.

P

Petrochemical Industry, 235.
Phosphate Class, 11, 56.
Phosphate Minerals, 56, 57.
Physical Properties of Minerals, 7.
Platinum, 13, 14, 16, 23, 27, 31, 38, 43, 47, 51, 57, 172, 177.
Portland Cement, 73, 150, 249, 251, 252, 253, 254, 255, 256.
Production, 27, 57, 89, 93, 94, 97, 104, 112, 116, 117, 119, 121, 133, 143, 147, 149, 163, 164, 166, 167, 168, 170, 172, 173, 175, 177, 194, 205, 206, 210, 212, 213, 215, 217, 221, 222, 224, 225, 229, 241, 242, 246, 249, 251, 253, 256, 257.

S

Safety Regulations, 148.
Sedimentary Rock, 5, 7.
Silicate Class, 9.
Sodium Chloride, 11, 27.
Specific Gravity, 9, 64, 68, 69, 70, 71, 74, 76, 94, 104, 114, 121, 128, 152, 155, 156, 158, 159, 160, 161, 180, 185, 187, 188, 189, 191, 192, 197, 199, 204.
Sulfate Class, 10.
Sulfate Minerals, 10, 49, 50, 106.
Sulfide Class, 11, 15.
Sulfide Minerals, 9, 10, 11, 15, 21, 192.
Sulfosalt Minerals, 21.

T

Titanium, 11, 14, 69, 71, 107, 114, 185, 187, 219, 245.
Treatment, 57, 92, 95, 104, 118, 119, 123, 150, 172, 224, 248.
Triggering Cancer Development, 137.

V

Vanadium, 11, 21, 68, 88, 89, 107, 114, 124, 226.

W

Water Pollution, 178.

□□□